Brain Imaging

Henry N. Wagner, Jr.

Brain Imaging:

The Chemistry of Mental Activity

 Springer

Henry N. Wagner, Jr., MD
Professor of Environmental Health Sciences
Johns Hopkins Bloomberg School of Public Health
Professor Emeritus of Medicine and Radiology
Johns Hopkins School of Medicine
Baltimore, MD, USA

ISBN 978-1-84882-922-0 (PB)
ISBN 978-1-84800-307-1 (HB)   e-ISBN 978-1-84800-308-8
DOI: 10.1007/978-1-84800-308-8

British Library Cataloguing in Publication Data

Library of Congress Control Number: 2008940826

Printed on acid-free paper

Springer Science + Business Media
springer.com

Dedication

To all the wonderful students, technologists and faculty in Nuclear Medicine at Johns Hopkins 1958–2008.

Foreword

A Grand Design

Imaging the human brain has captured the imagination of the general public more than almost any scientific advance of my recollection. The ability to visualize in great detail the soft tissue of the brain substance has revolutionized clinical neuroscience and permitted early and noninvasive diagnosis for all sorts of brain diseases. More fascinating has been the use of imaging techniques to elucidate brain function. Magnetic resonance imaging can be used to monitor general activity in different parts of the brain. For me, however, the use of radioactive isotopes to visualize metabolism in intact humans is in some ways more important. One underlying strategy reflects work of Louis Sokoloff at the National Institutes of Health (NIH). In the mid-1970s he assessed glucose use with the nonmetabolized derivative 2-deoxyglucose. With the advent of positron emission tomography (PET) scanning, positron-emitting derivatives of 2-deoxyglucose were used to evaluate glucose use in the human brain.

I remember well the excitement surrounding the early PET scanning with 2-deoxyglucose. Medical institutions throughout the country invested many millions of dollars in cyclotrons and PET scanners. Besides the expensive equipment, a staff of chemists was required to rapidly synthesize the short-lived positron emitting ligands, and physicists were necessary to nurture the temperamental cyclotrons. Accustomed to simple, inexpensive, low-maintenance biochemical technology, I regarded the 2-deoxyglucose PET scanning enterprise as too much fuss for too little payoff.

In the early 1970s, Henry Wagner came to visit with me. He wished to apply to the NIH for funding to purchase all the goodies necessary to commence PET scanning. Numerous institutions across the country were jumping on the bandwagon, all doing similar 2-deoxyglucose monitoring. Henry eschewed copycat science and sought something innovative. He had read of our research labeling receptors for opiates and neurotransmitter receptors by the reversible binding of small drug-like ligand molecules. He asked for help in formulating a proposal to the NIH to label appropriate drugs with positron emitting isotopes to monitor receptors in humans. I was fascinated by the challenge and eager to help. For the grant application, one of the principal tasks was calculating various aspects of drug disposition, which would influence whether enough receptor-bound drug could accumulate to permit imaging through the skull. My faculty colleague Michael Kuhar had collaborated with me some initial receptor studies. His bachelor degree was in physics and his Ph.D. was in biophysics, so Michael had the mathematical skills to address the underpinnings of a successful grant application.

The grant was submitted, funded, and PET labeling of receptors in humans worked. I recall vividly the ecstasy of Mike and Henry when the first images of dopamine receptors were obtained. Henry insisted on being the first experimental subject. This led to a peculiar observation. Dopamine receptors are most highly concentrated in the caudate nucleus that contains the terminals of the major dopamine pathway that degenerates in Parkinson's disease. Repeated images indicated a major discrepancy between the right and left caudate of Henry's brain. Might he have a selective aberration of dopamine receptors in one side of the brain? With subsequent imaging of other receptors, the abnormality persisted. Presumably, Henry's brain has compensated for the anomaly as he is a well coordinated individual.

I became personally involved with the PET scanning effort when it came time to image opiate receptors. Although the PET scanning team didn't really need me, Henry felt it appropriate to include me as a collaborator and participant in the initial identification of opiate receptors. As was usual, Henry was the initial experimental subject. I served as the "anesthesiologist." One of the critical tasks in imaging receptors is to evaluate non-specific binding, the portion of the administered drug that is not associated with the receptor but linked to other tissue components. This can be estimated by administering a nonradioactive drug that displaces the radioligand from its receptor. In this case, we used the classic opiate receptor antagonist naloxone. To ensure that enough naloxone was administered to displace the radioligand from opiate receptors, the pharmacologists recommended administering 100 times the naloxone dose typically given to a patient suffering from a heroin overdose. Some years ago, we had discovered that high concentrations of naloxone blocked receptors for the inhibitory neurotransmitter GABA and would thus be predicted to cause convulsions. I had long since forgotten about that study, but it all came back to me when I began pumping the naloxone into Henry. Instead of his usual carefree, joking demeanor, Henry became irritable, anxious, and shaky. I was terrified that we were going to trigger a convulsion. Fortunately, naloxone is very short lived. Henry's "panic" subsided within a couple of minutes. The imaging was highly successful and many studies of opiate receptors in humans ensued. I resolved never again to insinuate myself into execution of the clinical study without knowing an awful lot about what was going on.

I relate the above-mentioned anecdotes to convey the essence of Henry Wagner—a brilliant, incisive, creative, and audacious investigator. Although these elements are on display in this ambitious volume, rather than just providing reminiscences of his experience in using nuclear medicine to study the brain, he provides an overarching mantle in which to appreciate all the things we can learn from imaging the brain. He commences with an explanation of neurotransmitters with a special focus on serotonin, the neurotransmitter most linked to emotional behavior. He even provides a crash course in molecular biology and how genetic malfunctions lead to disease. He provides an overview of how mental function can be abstracted from the molecular interface of diverse neurotransmitters in discrete brain regions. Most importantly, he provides a sophisticated yet readily understood description of the underpinnings of radioactivity and how a series of investigators, going back to the days of Pierre and Marie Curie, worked out the use of radiolabeled chemicals as tracers. It has long been my feeling that too few people appreciate the critical role of radiotracers in the revolution in biochemistry and molecular biology that characterize the second half of the 20th century. Without such tracers it would never have been possible to map the major metabolic pathways in the body. The breakthroughs in molecular genetics derived heavily from the use of radio-isotopes. All of our own work identifying neurotransmitter receptors would never have been possible without the use of drugs and neurotransmitters labeled very selectively with isotopes.

Henry seamlessly integrates descriptions of diverse imaging techniques with a with range of mental states and diseases that have been investigated by PET scanning. He links these into a coherent view of the molecular underpinnings of brain function. For example, he describes his own research in 1950 on the effects of the drug bulbocapnine, which mimics the motor manifestations of catatonic schizophrenia. He then shows how cocaine, which enhances the effects of dopamine, can block the actions of bulbocapnine. He relates these findings to the catatonia of postencephalitic Parkinson's disease, which is dramatically reversed by the increased formation of dopamine after treatment with L-dopa. He links these observations to what he has learned about dopamine systems by using PET scanning.

Henry writes with a degree of lucidity and enthusiasm that renders the volume accessible both to the specialist and the educated lay public. We are in the debt to Henry Wagner for conveying to us his remarkable life's work and transforming insights.

Solomon H. Snyder
Baltimore, MD

Acknowledgement

To Judy Buchanan for her help in correcting the page proofs.

Contents

1

A Molecular Theory of Disease

The atom is the basic building block of matter. The model of the atom is a positively charged nucleus and surrounding negative electrons. The chemical behavior of an atom depends on the numbers of these electrons. Most atoms exist as part of molecules. All life on earth depends on molecular processes. Most of these processes need a molecular catalyst to speed up chemical reactions taking place at body temperature.

Molecules are the smallest parts of matter that retain the chemical properties of the substance. A molecule consists of at least two atoms held together by chemical bonds. Everything in the body is made of molecules. If the nucleus of an atom is very large, the atom is often unstable, undergoes radioactive decay, and emits smaller particles or high-energy photons. With special instruments, these particles can be detected from outside the body.

An example of a molecular disease is the panic state, which results from the flooding of the brain with epinephrine (adrenaline), a hormone carried in the blood. It is also a neurotransmitter released into neuronal synapses of activated neurons of the autonomic nervous system. Both epinephrine and norepinephrine are responsible for the emotions anger and fear.

Damage to the blood-brain barrier (BBB) as a result of trauma or disease permits hormones (molecules), such as epinephrine, or toxic substances to reach the brain. Normally, the BBB keeps the contents within blood vessels from reaching the brain. Closely packed endothelial cells line blood vessels. The endothelial cell membrane has transport systems that facilitate the movement of desired molecules and nutrients into the brain, while keeping undesirable molecules out. Damage to the BBB from trauma may be involved in posttraumatic stress disorder.

Epinephrine is released from the adrenal medulla by activation of the sympathetic nerves of the autonomic nervous system. Norepinephrine is released from the adrenal medulla. About 20% of the total catecholamine released from the adrenal is norepinephrine. Most norepinephrine released by sympathetic nerves is taken back up into presynaptic neurons. A small amount diffuses into the blood and circulates throughout the body. When the sympathetic nervous system is highly activated, the amount of norepinephrine entering the circulation increases.

Many drugs diffuse across the BBB or enter the brain via active transport. More people take drugs to affect mental activities than to fight infection. Drugs can produce molecular diseases of the brain just as antibiotics can result in the growth of virulent pathogens. Such drugs can disturb brain chemistry, just as antibiotics can increase the number of antibiotic-resistant bacteria. When abused, psychotropic drugs can provoke harmful molecular responses in brain chemistry. Harmless environmental events can become to be viewed as threats, making the person fearful, aggressive, and violent.

There is a molecular theory of mental disease, based on four postulates: (1) An external or internal event or threat brings about the release of specific hormones and/or neurotransmitters. (2) These produce an emotional response. (3) The emotional response affects thoughts. (4) The emotions and thoughts lead to actions. Molecular processes in a disease can be thought of as a stream, running from threats to emotions, from emotions to thoughts, and from thoughts to actions. The fight-or-flight response is an example.

Consciousness is self-awareness, which emerges as sensations become perceptions. These sensations elicit emotions, followed by thoughts and actions. Consciousness is the result of the molecular and electrical processes in the brain, underlying the integrated activity of billions of neurons that bring about self-awareness, emotions, thoughts, and ideas. Sensations become perceptions when integrated with past experiences. Consciousness is to the brain what walking is to the legs, or seeing is to the eyes.

> "Everyone knows what attention is. It is the taking possession by the mind, in clear and
> vivid form, of one out of what seem several simultaneously possible objects or trains of
> thought. Focalization, concentration, and consciousness are of its essence" (James, 1890).

Emotions evolved in hunter–gatherers long before the evolution of language and reasoning. Animals and early human beings often experienced fear, causing them to flee. They became angry as they faced the sources of danger.

The autonomic nervous system evolved before reason and other higher brain functions and plays a role in emotions that help people survive. Parkinson's disease affect the autonomic nervous system, characterized by orthostatic hypotension and impaired sweating.

1.1. Fight-or-Flight Response

Hans Selye of the University of Montreal (Montreal, QC, Canada) proposed that the body has a uniform response to stress, whether resulting from strenuous muscular activity, prolonged food deprivation, or exposure to toxins. Pituitary hormones, steroid hormones, and adrenal catecholamines are secreted in large quantities in the general adaptation syndrome. The body goes through three sequential stages: (1) an alarm reaction, during which the body prepares for fight or flight; (2) resistance, during which biological changes counteract the deleterious effects caused by the stressing agent; and (3) exhaustion.

During the fight-or-flight response, the amygdalas activate the hypothalamus, which activates the adrenal glands to produce epinephrine and norepinephrine. The amydalas are part of the limbic system lying just beneath the thalamus, and they include the hypothalamus and hippocampus. The limbic system includes the basal ganglia, regions of the brain involved in movement and memory as well as emotions. People remember emotionally charged events.

During stress, the sympathetic and parasympathetic components of the autonomic nervous system increase the rate and force of contractions of the heart. Adrenergic neurons in the left stellate ganglia innervate the right ventricle; the right stellate ganglia innervate the anterior and lateral portions of the ventricles.

Even in the absence of stress, norepinephrine affects mood, sleep, and alertness, being at its lowest blood levels during sleep. People can develop an addiction to their own epinephrine and norepinephrine, and take pleasure in activities such as sky-diving, motor cycling, or mountain climbing to produce an adrenaline rush. They can also become addicted to endogenous enkephalins (endorphins) which have opiate-like effects, inducing pleasure and feelings of well being.

The neurotransmitter serotonin is released by strong emotions. In 1957, Julius Axelrod at the National Institutes of Health (Bethesda, MD), found that the enzyme monoamine oxidase (MAO) inactivates norepinephrine and other monoamines. In 1961, he found that neurotransmitters are removed from synapses by being taken back up into the presynaptic neurons that had secreted them, and are secreted again during later neuronal events. Drugs that block MAO activity prevent the breakdown of the monoamines, resulting in the storage of large amounts of the neurotransmitters in presynaptic vesicles until they are ready to be released again.

Serotonin (5-hydroxytryptamine; 5-HT) is involved in the regulation of the cardiovascular and respiratory systems, sleep, aggression, sexual behavior, food intake, anxiety, mood, motor output, neuroendocrine secretion, nociception, and analgesia.

Serotonin affects motor activity by inhibiting the impact of sensory information and by coordinating autonomic and neuroendocrine function during demanding motor activity. Terminals of neurons secreting 5-HT are located in the primary and secondary motor areas of the brain.

Serotonin plays a role in the modulation of social behavior and in reward processing. Deficiency of tryptophan, the serotonin precursor, leads to reduction in serotonin production, and brings about reductions in the level of cooperation shown by participants in an experimental game-playing experiment. Serotonin also plays a role in socially cooperative behavior (Wood, 2006).

The hormone melatonin induces sleep, but serotonin is involved in dreaming; both are produced in the pineal gland. If the brain has low serotonin levels, dreaming will not occur. Serotonin is also important in learning, memory, and mood. It is deficient in the brains of depressed patients. Serotonin deficiency can result in an inability to fall asleep at night, panic attacks, loss of concentration, and thoughts of suicide or attempted suicide.

Melatonin (5-methoxy-N-acetyltryptamine) is produced from tryptophan by the pineal gland, retina, and gastrointestinal tract, and it is bound by melatonin receptors. Nobel Prize laureate Julius Axelrod discovered the role of melatonin and the pineal gland in regulating sleep–wake cycles (circadian rhythms). Production of melatonin by the pineal gland is stimulated by darkness and inhibited by light. Melatonin inhibits secretion of luteinizing hormone and follicle-stimulating hormone from the anterior pituitary gland.

Alcohol lowers levels of serotonin in the brain, and many alcoholics have dreamless sleep, devoid of rapid eye movement activity. When alcoholics withdraw from alcohol, many experience delirium tremens (DTs), manifest by shaking, sweating profusely, anxiety, and hallucinations. Alcohol depletes the brain of serotonin, the levels of which may rise to higher than normal levels with the withdrawal of alcohol. Excessive production of serotonin is thought to cause the hallucinations, which characterize DTs.

Norepinephrine is synthesized from tyrosine, stored in presynaptic vesicles, and released from vesicles upon the arrival of axonal action potentials. After its release into synapses, unbound norepinephrine is taken back into presynaptic neurons by norepinephrine transporters, which terminate its effect. The reuptake process clears 70–90%

of the norepinephrine released into synapses. Norepinephrine remaining free in the synaptic cleft is metabolized by two monoamine oxide enzymes, MAO A and B.

Serotonin has an opposite effect from norepinephrine and epinephrine. In the face of a threatening environment, the serotonergic system increases feelings of well being, calm, relaxation, confidence, and concentration. Serotonergic neuronal networks counterbalance the effects of noradrenaline associated with arousal, fear, anger, tension, aggression, violence, obsessive-compulsive actions, overeating, and anxiety and sleep disturbances.

Serotonin, first isolated in 1948 by Maurice M. Rapport, was initially thought to be only a vasoconstrictor. The word *serotonin* implied a serum agent identified as a vasoconstrictor. The body of the average adult human possesses 5–10 mg of serotonin, 90% of which is in the intestine, and the rest is in blood platelets and the brain.

Serotonin was subsequently found to be a neurotransmitter with several subtypes of receptors: 5-HT1, 5-HT2, 5-HT3, 5-HT4, 5-HT5, 5-HT6, and 5-HT7. The physiological function of each receptor subtype has not been established. With the exception of the 5-HT3 receptor, which is a ligand-gated ion channel related to N-methyl-D-aspartate (NMDA), GABA, and nicotinic receptors, all of the 5-HT receptor subtypes belong to the group of G protein-linked receptors. Specific agonists and antagonists for each receptor system are the focus of new drug development.

A tryptophan-free diet results in a fall in brain serotonin levels after 4 hours. Lacking the inhibitory regulation of serotonin, neuronal activity is insufficiently modulated, resulting in the person becoming angry, depressed and impulsive, losing control of emotions.

In August 2004, fluoxetine hydrochloride (Prozac), a selective serotonin reuptake inhibitor (SSRI), was found in drinking water in the United Kingdom. How it got there, no one knows. Is it possible that *mind-altering* drugs might someday be added to drinking water the same way that fluoride is added to prevent caries?

Prozac is the most frequently prescribed drug in the United States. In Listening to Prozac (Knopf), published in 1993, and Against Depression (Viking), published in 2005, Peter D. Kramer discusses drugs, called SSRIs that are widely used to treat depression. They inhibit the reuptake of synaptic serotonin into presynaptic neurons. They have helped many persons, but their use also has created many problems. In *Forbes* magazine (November 2004), the following statement was made: "The 1990s made pill-popping for happiness an acceptable therapeutic alternative for millions of even mildly depressed patients."

Placebo controlled trials in children and adults showed that 60 of 9,219 (0.65%) patients given the SSRI drug paroxetine, compared with 20 of 6,455 given placebo (0.31%), had what was called a *hostility event*, that is, commiting violent acts.

When monkeys have low serotonin levels, they become aggressive. This response is thought to be a model for violence against oneself or family members, possibly related to a failure to counteract epinephrine effects. High aggressiveness in monkeys was correlated with low levels of metabolic products of serotonin in spinal fluid. In humans, low levels of serotonin metabolites in spinal fluid also correlate with the severity of physical aggression.

As Peter Kramer has written, "Assertiveness gets you what you need, and correlates with high brain-serotonin levels. Uncontrolled violence against oneself or others correlates with low brain-serotonin levels."

SSRIs are involved in the actions of the three best known neurotransmitters–serotonin, norepinephrine, and dopamine–monoamines metabolized by the enzyme MAO. MAO inhibitors inhibit SSRI action. These three amines have been found to be low in the brains of patients who are clinically depressed. SSRIs prevent the breakdown of these mono-amines. Persons beginning to take Prozac may experience epinephrine-related effects after 3 or 4 days, perhaps a reaction to the transiently high levels of serotonin. On March 22, 2004, the Food and Drug Administration warned that SSRIs might cause anxiety, agitation, panic attacks, insomnia, irritability, hostility, impulsiveness, extreme restlessness, hypomania, and mania.

Serotonin has been implicated in cognition, affecting memory, perception and attention, mood, aggression, sexual drive, appetite, energy level, pain sensitivity, endocrine function, and sleep. Patients with schizophrenia are thought to have abnormal dopaminergic and serotonergic neurotransmission. Their cerebral ventricles may be enlarged, and their temporal lobe volume decreased, but such structural manifestations are not specific. The focus was initially on dopamine.

The NMDA hypothesis, called the glutamatergic dysfunction hypothesis, of schizophrenia is based on the action of glutamate on NMDA receptors on GABAergic, serotonergic, and noradrenergic neurons that inhibit two major excitatory pathways in the retrosplenial cortical neurons. (Coyle, 1996).

Serotonin receptors are decreased in the cerebral cortex and striatum in aging persons. Serotonin receptors are increased in suicide victims with major depression. Patients with chronic fatigue syndrome have decreased numbers of serotonin transporters in the anterior cingulate (Watanabe et al., Mind/Brain Symposium, 2002). Drugs used to treat patients with depression increase the levels of serotonin and noradrenaline in the brain.

Schizophrenia is manifest by impaired cognition and failure of appreciation of reality. Patients with schizophrenia have positive symptoms, such as hallucinations, mood disorders, or negative symptoms, such as impaired interpersonal relations, social withdrawal, and difficulty in holding a job. Positive symptoms are related to decreased dopaminergic activity. Negative symptoms, such as catatonia, are related to serotonergic activity.

Clozapine and other antagonists of serotonin and D2 dopamine receptors, including risperidone, olanzapine, sertindole, ziprasidone, and quetiapine, improve patients because of their effects on several brain regions (Bryan L. Roth and Herbert Y. Meltzer, The Role of Serotonin in Schizophrenia, Neuropsychopharmacology).

Wooley, Shaw, and Gaddam first advanced the hypothesis that decreased serotonergic activity was the cause of schizophrenia. Research over the next 25 years did not support this hypothesis, even though there were decreased densities of specific types of 5-HT receptors in the brains of schizophrenic patients.

Interest in the role of serotonin in schizophrenia increased when numerous 5-HT receptor subtypes were found. The extraordinary ability of the drug clozapine to improve patients with schizophrenia was attributed to its ability to block 5-HT2A receptors.

Serotonin has been linked to aggressive behavior as a result of activation of the limbic system (olfactory bulb, amygdala, and hypothalamus), the prefrontal cortex, and a region of the periaqueductal gray (Wood et al., 2006).

Tiihonen et al. (1997) examined the serotonergic system with single photon emission computed tomography in 52 subjects (21 impulsive violent offenders, 21 age- and sex-matched healthy controls, and 10 nonviolent alcoholic controls. The specific binding of iodine-123-β-carbomethoxy-3(4-iodophenyl) tropane to serotonin transporters in the midbrain of violent offenders was lower than that in the healthy control subjects, thought to be due to occupation of the serotonin transporters by increased amounts of synaptic serotonin or to an impairment or absence of the presynaptic serotonergic neurons themselves.

Serotonin deficiency is related to a broad array of emotional and behavioral problems, ranging from depression, premenstrual syndrome, anxiety, alcoholism, insomnia, violence, aggression, suicide, and compulsive gambling.

2

The Master Molecule

The genome in the nucleus of every human cell (except red blood cells) consists of 6 billion deoxyribonucleic acid (DNA) nucleotides packaged in 48 chromosomes. The life of every human being begins with DNA, a polymer of a long series of nucleotides, with a backbone of five-carbon sugars. Ribonucleic acid (RNA) contains ribose in the place of deoxyribose. Human DNA contains 22,000 genes, containing 750 MB of data (the Bible contains 5 MB).

Humans and chimps evolved separately from a common ancestor about 6 million years ago. Nearly 99% of the gene sequences in chimps are identical to those of humans. Macaque monkeys branched off the family tree of a common ancestor about 25 million years ago. Even in 93% of macaque DNA is the same as that of human beings. Ninety-nine percent of each person's DNA is identical to that of all other human beings.

Today, the evolution of genes, programmed cell death (apoptosis), and the action of messenger RNA (mRNA) are three major targets of research. mRNA contains the blueprint for every protein in the body. It is transcribed from a DNA template, and carries information to ribosomes, the sites of protein synthesis. The sequences of nucleic acid polymers are translated by transfer RNA (tRNA) into amino acid polymers. tRNA recognizes the three-nucleotide sequences that encode each amino acid. Ribosomal RNA directs the ribosome's production of proteins. Codons carry the messages that terminate protein synthesis.

The DNA of genes is made of four nucleotide bases: two purines, adenine (A) and guanine (G); and two pyrimidines, thymine (T) and cytosine (C). The genetic code is based on these four letters, AGCT, that encode the amino acids making up the body's peptides and proteins. The genetic code is the same in all living creatures.

Nobel Prize winner Sydney Brenner showed that a sequence of three nucleotide bases in a row encode each specific amino acid. RNA strands complementary to a DNA strand are responsible for the process called transcription, which is brought about by the action of the enzyme, RNA polymerase.

DNA not only passes information from one generation to the next, but throughout a person's entire life it is involved in the continual reproduction of the molecules in all the cells of the body. From the moment of conception until the time of death, it is the continual transfer and selective expression of genetic information that makes life possible.

The total complement of genes in a living organism is called the genome. Three billion nucleotides that make up human DNA carry instructions that control the structure and function of all cells of the body (except red blood cells, which have no nuclei). Three percent of human DNA encodes information. The remaining 97% is called junk DNA, the function of which is not yet known.

Specific genes, called exons, are activated by promoter genes located along the sequence of the DNA molecule. Noncoding sequences are called introns, and do not lead to the production of products. Exons encode specific peptides and proteins: structural molecules, neurotransmitters, and enzymes regulating chemical processes.

The evolution of different species is the result of genetic mutations, which occur in organisms as primitive as bacteria. Bacteria can exchange genes. When grown in a culture medium, they reproduce every 20 or 30 min. Mutations enhance their survival in the face of a harmful environment, such as an environment containing antibiotics. Mutations result in the reproduction of bacteria resistant to the effects of antibiotics. The antibiotic-resistant bacteria survive and reproduce, creating a new strain of bacteria.

The nervous system lies between genes and behavior. In his Nobel Prize lecture in 2002, geneticist Sydney Brenner said, "Behavior is the result of a complex, ill-understood set of computations performed by nervous systems. It seems essential to decompose the question into two: one concerned with the question of the genetic specification of (different) nervous systems and the other with the way nervous systems work to produce behavior ... We are drowning in a sea of data and starving for knowledge.

"The biological sciences have exploded, largely through our unprecedented power to accumulate descriptive facts. How to understand genomes and how to use them is going to be a central task of our research for the future. We need to turn data into knowledge. Biology has become so focused on genetics that we have forgotten that the real units of function and structure in an organism are cells and not genes ...

After the basic principles of information transfer from genes to proteins had been established with the identification of messenger RNA, the discovery of the mechanism of protein synthesis and the structure of genetic code, it was natural for some of us to ask whether the lessons learnt in molecular biology could be applied to the genetics of more complex phenotypes."

Brenner advises students to study the relationship of genes to biological processes: "All you have to do is to find a gene and have it sequenced and then make some protein using the gene and get someone to determine its amino acid sequence."

One can use molecular imaging with radioactive tracers to determine the role of proteins and other molecules in the process of living. More recently, optical imaging, based on luminescence or fluorescence, also makes this possible.

Tracers emit radiant energy as radioactive particles or photons, which can penetrate the body to be detected by imaging devices that show where the source of radiation lies within the body.

Molecular imaging makes it possible to detect abnormalities of one or more modifiable molecular manifestations associated with abnormal perceptions, thinking, and behavior. Measuring these manifestations of disease provides the foundation of molecular medicine.

3

Genes and Disease

The struggle for existence continues at whole body, cellular, and molecular levels. Plant and animal species survive by producing the most offspring. Plants launch thousands of seeds. Some have stickers that attach themselves to the feet and legs of passing animals, making it possible for them to survive and reproduce.

Genes are DNA sequences that are copied by the action of the enzyme RNA polymerase, and are passed on by transfer RNA to ribosomes, where proteins are synthesized in a process called translation. Ribosomes assemble amino acids into polypeptide chains; messenger RNA provides a template to join together the correct sequence of amino acids.

DNA and RNA can be radiolabeled with carbon-14, hydrogen-3, or phosphorus-32 and then separated into their constituent nucleotide sequences by enzymatic digestion. Restriction enzymes cut double-stranded DNA through each of the sugar-phosphate backbones, without damaging the nucleotide bases.

In the millions of nucleotides in the segments that make up molecules of DNA, there are millions of differences in the nucleotide sequences between any two persons. It is possible to determine which segments are inherited from which parent. Some fragments have been correlated with specific diseases in a process called genetic linkage.

Genetic linkage maps show the relative locations of specific markers along a chromosome. A marker may be related to a trait, such as the Lesch–Nyhan syndrome. Restriction fragment length polymorphisms and specific DNA sequences, called sequence-tagged sites, are markers that make it possible to identify persons with a specific disease.

A person's genome reflects the history of the human race, the specific population from which the person descended on both parental sides, and mutations that have occurred over the course of the person's life. These mutations are responsible to a large degree for the differences among persons, including differences in the risk of certain diseases.

The first linking of genes to a specific human disease was in 1908, when Archibald Garrod linked the disease alkaptonuria to a hereditary deficiency of the enzyme homogentisate 1,2-dioxygenase that degrades the amino acid tyrosine. A single mutated gene results in a failure of this process, and results in the disease. His finding of a relationship between a gene and a disease led to the search for other inborn errors of metabolism, and the foundation of molecular genetics. Color blindness and sickle cell disease are other examples of diseases related to a single gene abnormality.

In November 1949, chemist Linus Pauling and colleagues showed that an abnormal form of hemoglobin caused the disease sickle cell anemia. People with sickle cell anemia have two copies of the abnormal hemoglobin gene, one copy inherited from each parent. People who inherit a sickle cell gene from one parent and a normal gene from the other have sickle cell trait. Their hemoglobin is normal, but they can pass the abnormal gene on to their children. Another example of a genetic linkage study was the discovery of the gene related to Huntington's disease in the 1980s.

The hunt for genes related to disease has become a major industry. Many of the abnormal genes are oncogenes that lead to uncontrolled cell growth. Television ads encourage women to "contact their physician or call toll-free numbers" to find out whether they carry any genes that increase their risk of developing breast or ovarian cancer. The tests cost >$3,000.

Mutations of the genes BRCA1 and BRCA2 increase the risk of the subsequent occurrence of breast cancer by 35–84% before the woman reaches the age of 70, and a 10–50% increase in the probability of developing ovarian cancer. Mutations in these genes account for less than 10% of all patients who develop breast cancer. Only 1 in 400 women in general have these mutated genes, so the cost of genetic testing to find a single woman predisposed to breast cancer would be very great if one contemplates population screening.

The Americans with Disabilities Act permits an employer to require prospective employees to take a genetic test as a condition of employment, just as they permit requiring a physical examination for certain jobs. Present laws do not prevent private insurance companies from denying insurance, or setting high premiums on the basis of genetic information, but 47 states offer some degree of protection from insurance discrimination.

In 2007, a study of 1,199 Americans by the Genetic and Public Policy Institution of Johns Hopkins found that 86% of the participants trusted their doctors to have access to the results of their genetic testing. The others were unwilling to have genetic tests because they feared discrimination in seeking jobs.

Although environmental factors, including economic, social, and educational status, play an important role in most diseases, single gene mutations are found in approximately 6,000 diseases. These diseases are called monogenic diseases, in which a single mutated gene can result in the disease. Heart disease, diabetes, Alzheimer's disease, psychiatric disorders, and osteoarthritis do not result from a change in one or even a few genes, but from an interaction of between the environment and a large number of genes.

Preimplantation genetic diagnosis can detect abnormal genes, such as those related to Huntington's disease, certain types of deafness, Down's syndrome, cystic fibrosis, sickle cell disease, or breast and colon cancer in human preembryos before they are implanted in the uterus in the process of in vitro fertilization. Prospective parents concerned about passing a serious gene-based disease to their child often take genetic tests, especially if one or both come from a population known to have a high incidence of gene-related disorders. If an embryo is found to be genetically defective, it can be destroyed.

In 1980, Botstein, White, Skolnick, and Davis developed a method for creating a complete genetic linkage map. Radiolabeled gene segments can be detected in extremely small samples, even in single cells. In humans, there are 60 trillion cells with 23,000 segments of the DNA molecules that control protein synthesis.

Starting with a simpler living system, Brenner and his colleagues are examining the gene segments in each of the 300 neurons of the worm, *Caenorhabditis elegans*, to try to construct models of its neuronal networks.

Fruit flies (Drosophila), which have played a major role in genetic research, have 10,000 neurons, and they will be the first focus, followed by the examination of the neuronal networks of the human nervous system as synapses come and go throughout life.

Genes determine the differentiation and growth of the cells of the body. Examining them is becoming an important part of today's health care. Many persons want to determine the risk of developing certain diseases during their life. Problems can arise if the tests are advertised and sold directly to the public. Yet, many are willing to send in a check and a bit of saliva to assess their risks of future disease. The tests are often marketed commercially on the basis of unproved, ambiguous, false, or misleading claims.

The results of the tests often form the basis of important decisions. For example, a woman with a family history of breast cancer may take the genetic test to decide whether to have surgery to remove a breast or ovaries to try to prevent cancer in the future. An inaccurate test may lead to risky, unnecessary treatments.

Human decisions are often made without considering probabilities. In 2002, Daniel Kahneman shared the Nobel Prize in Economics by examining how decisions are made. The way a problem is formulated influences a decision. When offered a choice formulated one way, the person may display risk-aversion behavior. When the choice is formulated in a different way, the person may be willing to accept the risk. If they consider probabilities, people often make decisions based on a subjective assessment of probabilities which may not be accurate. An asymptomatic person who receives the result of a genetic or biochemical profile that predicts an increased risk of disease may change life style, avoid environmental or occupational hazards, try to avoid emotional stress, or change diet. Early recognition of predisposition to Alzheimer's disease can motivate a person to increase mental activities, such as solving crossword puzzles. Recognition of a predisposition to osteoporosis can lead persons to increase calcium in their diet and exercise more.

Phenotypes can be related to genotypes by molecular imaging. One can examine the role that specific genes play in homeostatic processes. Regional biochemical processes must be accurately quantified to correlate biochemical defects with genes and the clinical manifestations of a disease. Radioactive tracers, and more recently, optical tracers, can be used together with perturbations, such as administering a drug.

Brain size and chemistry are strongly influenced by genes. Language and other cognitive functions evolved over millions of years, while the size of human brain tripled. Microcephaly is a disease characterized by a 70% reduction in brain size. It results from a mutation of the spindle-like microcephaly associated (ASPM) gene (Lahn, 2005). There is a failure of separation of chromosomes during meiosis, the process where one diploid eukaryotic cell divides into four haploid cells, called gametes. Persons with microcephaly have a small cerebral cortex and often suffer from mental retardation.

The evolution of large brains helped humans outwit or escape from predators, learn to use fire, and develop communal living. Strong social bonds, high levels of intelligence, intense parenting skills, and long periods of learning made them better able to deal with hostile environments. How these factors were influenced by an increased size of the brain is conjectural. Early humans lived in open and dry areas, where it was difficult to escape from predators or find water. Those with larger brains probably mated earlier, lived longer, and bore more children who were more likely to survive and bear their own children.

Neurologist Christopher Walsh of Harvard Medical School and Geoff Woods at the University of Leeds share the belief that humans developed large frontal lobes that increased social interactions, language, and cognition. Yerkes, Hammock, and Young of the Department of Psychiatry and Behavioral Sciences at Emory implicate the gene expressing the antidiuretic hormone as affecting social behavior, a gene related to social behavior in many species (Yerkes, et al Nature, June 17, 2004).

Throughout life, genes are selectively turned on and off as they regulate neuronal networks and biochemical processes in the brain. Genes provide the blueprints throughout life, as forces in the environment determine which genes are turned on or off. Genes encode phenotypes, including thinking, feeling, and behavior, by activating or inhibiting hormones, neurotransmitters, and neuronal networks.

Shakespeare said that the brain provides hope during childhood, and wisdom in old age. Polonius said that he had found the cause of Hamlet's lunacy. Not knowing whom to trust, he lashed out at everyone around him.

"The mind is but a barren soil, a soil which is soon exhausted, and will produce no crop, or only one, unless it be continually fertilized and enriched with foreign matter" (Reynolds, 1772). The genome is the foundation of a person's vulnerabilities, strengths, and potentialities.

Interacting with environmental and cultural forces, the genome programs a person's phenotypes throughout life. Nature and nurture are joined throughout life as are our right and left hands. Genotypes and phenotypes are the fingers of those hands, creating ideas, thoughts, and actions. Raymond Cattell estimates that 80% of human intelligence and 30% of emotional traits are strongly influenced by genetic factors.

Our primate ancestors lived in social hierarchies with clear dominance relationships (Gardner, 1995). "Primates recognize individual members of their species from an early age, compete with one another for positions within the hierarchy, and ultimately assume specific relationships of dominance or submission to members of their own species." One need only watch chickens feeding for a few minutes to see an example of a pecking order.

Humans learn a lot by imitating others. "The decision about which models to imitate is crucial as a child grows up. Size, strength, skill, intelligence, attractiveness, and gender all contribute to the determination of which organisms will occupy superior positions in the emerging social hierarchy. Researchers studying early socialization of children have documented the importance of the establishment in early life of a strong and secure bond of attachment between infant and caretaker" (Howard Gardner).

People who have serious difficulty recognizing faces suffer from a disease called congenital prosopagnosia (CP), the result of a genetic abnormality. Such relationships support the idea that some cognitive abilities are determined genetically (Behrmann and Avidan, 2005). Persons with the disease CP cannot discriminate between two faces presented side by side. Functional magnetic resonance imaging studies have shown that the regions of the brain activated when normal persons recognize faces are not activated in persons with CP.

3.1 The Human Genome Project

To help understand better how radiation damages genes, the U.S. Department of Energy began a landmark project in the 1980s to try to identify and sequence all 34,000 genes in a human. Charles DeLisi had worked on computational models in biology at the National Cancer Institute and developed methods for gene sequencing, in collaboration with scientists at the Los Alamos National Laboratory. He was a project leader in biological research at the Department of Energy (DOE), and he proposed that the DOE institute a Human Genome Project. In 1987, he persuaded the DOE to provide $5.5 million for this program. In 1988, due to the influence of Senator Pete Domenici of New Mexico, the program was finally begun. By June 20, 2000, it was possible to publish the first draft of the human genome. It has become a major factor in the design and development of a whole host of new pharmaceuticals by linking genes to diseases (genetic linkage). Gene products provide targets for drugs that stimulate or suppress molecular processes.

Diseases result from genetic factors and from the deleterious effects of things in the environment, from microbes, trauma, violence, or deficient diets. They often involve abnormalities in molecular processes affected by both genetic and environmental factors. Diseases, including heart disease, diabetes, cancer, mental illnesses, and age-related disorders, are complex, influenced by many different genes, that affect many processes.

Different breeds of dogs domesticated 15,000 years ago are susceptible to specific diseases. The Samoyed breed is at high risk of developing diabetes. The Rottweiler breed develops osteosarcomas. Springer spaniels are at a high risk of developing epilepsy. Doberman pinschers are prone to develop narcolepsy. The mutated genes in these animals are identical to those found in humans and in other species.

The Human Genome Project made possible the sequencing of the genomes of bacteria, yeast, fruit flies, mice, rats, and other plants and animals. The fruit fly has 14,000 genes; the worm *C. elegans*, 19,000 genes; yeast, 6,000 genes; and *Escherichia coli*, 4,300 genes. Many genes are the same in organisms as different as fruit flies, yeast, bacteria, and plants. The average gene in humans is a string of 100,000 base pairs. Humans and mice have 200 genes in the same sequence in their chromosomes.

Examples of genes linked to traits that predispose a person to different diseases include the following:

1. TNFS -4: myocardial infarction
2. NOS3: myocardial infarction
3. MMP3: myocardial infarction
4. KL: stroke, coronary artery disease
5. GNB3: hypertension, obesity
6. LCT: lactose intolerance
7. PER-2: sleep disturbances
8. SCL6A3: substance abuse
9. OMT: alcoholism

10. CHRNA6: tobacco addiction
11. CHRNA4: low risk of tobacco addiction
12. DRD-4: aggressive personality
13. MAOA: aggression and violence
14. A2: blue eyes, fair skin
15. APOE: Alzheimer's disease

3.2. Genetic Linkage Analysis

It is not necessary and would be enormously expensive to obtain a person's entire genome. Single-nucleotide polymorphisms (SNPs; snips) are specific DNA sequences. Single nucleotides—A, T, C, or G—are what make up the genes. For example, two DNA sequences, such as *AAGCCTA* or *AAGCTTA*, differ only in a single nucleotide.

Microarrays are commercially available that contain a million SPNs that can be searched for specific gene sequences. Affymetrix (Santa Clara, CA) has developed a gene chip that can examine automatically a million SNPs at one time. SNPs do not cause disease, but they can identify segments of mutated genes that increase the risk of developing a disease. Statistical analysis is used to detect the mutated genes.

The National Institutes of Health (NIH) makes available to the public a web-based database of large population studies, beginning with the Framingham Heart Study, which has studied 9,000 participants over three generations and found 550,000 genetic variants by SNP analysis.

Gene abnormalities include indels, in which a single DNA segment has either been added or deleted from a gene. There may be multiple copies of what normally should be a single gene; or a single gene may be missing from a chromosome or reinserted backwards.

Ninety-nine and a half percent of their DNA is the same in any two persons. The 0.5% differences are responsible for the differences in phenotypes. Craig Venter's genome was published in the journal *PLoS Biology*. He identified DNA variants in his own genome that increase the risk of alcoholism, coronary artery disease, obesity, Alzheimer's disease, antisocial behavior, and conduct disorder, as well as genes that decrease those risks. With this knowledge, he began taking anticholesterol drugs.

DeCODE is a biopharmaceutical company based in Reykjavik, Iceland, that uses genetic technology to develop therapeutic drugs and diagnostic agents. The company has the unique resource because it has the records of the entire population of Iceland, living or dead. The ancestors of today's population are known with great accuracy, and their medical records are excellent. Most of the population has provided their DNA samples. DeCODE is trying to understand the genetic underpinnings of disease.

The company markets a test for the gene TCFL2. If an individual has two copies of this gene, one from each parent, it doubles the risk of that person developing type 2 diabetes. If there are no other risk factors for diabetes, the chance of getting the disease is so low that it is not worth taking the test.

The cost of a person's having his own entire genome analyzed is falling, but it is extremely expensive. The extraction of the first genome in the Human Genome Project that was completed in 2003 cost $3 billion. Soon, the cost of one person's genome will be <$10,000.

More than 350 biotechnology companies are in Maryland, having been created by $38 billion invested soon after the White House announced the completion of the first draft of the human genome. By 2006, the investment dropped dramatically, because the benefits of deciphering came more slowly than enthusiasts promised when the project started.

Gene Logic, Inc. (Gaithersburg, MD) is a typical company founded to use gene technology to develop new drugs. The shares of the company opened at $42 on June 26, 2000, but dropped to a closing price of $1.30 only 6 years later. The new companies had planned on huge profits by selling parts of the genetic code to drug companies.

Other companies promote genetic tests on saliva to determine a person's risk of developing several diseases. It is not yet known how people will react to knowing their risks of specific diseases. Some believe that they will be able to make life style changes that will help them avoid or delay the onset of these diseases. They are encouraged to take tests that will tell them their risk of developing diabetes, obesity, prostate cancer, breast cancer, and glaucoma.

Examples of commercially available tests are those that will detect mutations BRCA1 and BRCA2, which are linked to the risk of breast cancer. Women carrying mutations of either the BRCA1 or BRCA2 gene have a risk of developing breast cancer by the age of 70. The commonly cited risk is of 50–80%. (Chase, 2008).

Many are trying to make the tests less expensive and more accurate in their predictions. Most tests are based on detecting snips that can detect sites on the genome where a single unit of DNA is abnormal.

A company, called Celera, went public. Its stock traded later at a high of $247. In 2006, Celera left the gene database field and focused on creating diagnostic devices for discovering interesting proteins as targets for drug development.

Some believed that gene-oriented companies would make molecular imaging obsolete, but this outcome has not proved to be the case. Positron emission tomography and single photon emission computed tomography are thriving.

Alan Guttmacher of the National Human Genome Research Institute, NIH, said recently: "One needs to separate the genomics *hope* from the genomic *hype*. Clearly, there was a lot of hype around the time of the (human genome) project."

3.3. The HapMap Project

The International HapMap Project is a partnership of scientists and research funding agencies from Canada, China, Japan, Nigeria, the United Kingdom, and the United States with the goal of helping researchers identify genes associated with human diseases and their ability to predict the patient's response to treatment.

In 2005, the first phase was completed with the identification of >50 disease-associated genes. On October 17, 2007, the second phase linked type 2 diabetes, Crohn's disease, elevated blood cholesterol, rheumatoid arthritis, multiple sclerosis, and prostate cancer to mutated genes. The project has catalogued >2.8 million SNPs in individual human genes.

3.4. Genetic Engineering

Genetic engineering is the manipulation of an organism's DNA. In 1976, with venture capitalist Robert Swanson, Herbert Boyer founded a company called Genentech to use recombinant DNA technology to produce medicines. In 1978, the company successfully inserted the human insulin gene into *E. coli* bacteria and then grew them to produce commercial amounts of insulin.

Subsequently, the company produced Rituxan for the treatment of patients with low-grade or follicular B-cell non-Hodgkin's lymphoma, which expresses the specific antigen CD-29. The company also produces Avastin, an antivascular endothelial growth factor, for treatment of patients with metastatic cancer of the colon or rectum; and Herceptin, a humanized anti-HER2 antibody for treating patients with metastatic breast cancer.

The way the process works is to select a gene that will lead to the production of the desired protein. Bacteria are then infected with the gene to produce large amounts of the therapeutic proteins that the gene specifies. The process is called synthetic biology, and it can be used to produce large quantities of specific proteins. Since 1999, the German company GeneArt has produced >15,000 genes for this purpose and sells them worldwide.

Genetic engineering is used not only for drug development but also for treatment of specific patients. In 1998, James Thompson of the University of Wisconsin isolated human embryonic stem cells, called pluripotent, because they can produce specific types of cells within the living body. This insertion of genes into the patient's body may someday be used to treat diabetes, rheumatoid arthritis, cancer, Parkinson's, and other diseases.

In 2007, Mario R. Capecchi, Oliver Smithies, and Martin Evans were awarded the Nobel Prize because of their creation of transgenic mice in which operating genes from other organisms have been transplanted. In a process called embryonic stem cell transplantation, defective genes are inserted that knock out the normal genes. Observing the resulting functional deficiencies tells what was the role of the knocked out gene. Today, transgenic mice are used widely for research on many human diseases.

Embryonic stem cell transplantation may be able to treat patients with Parkinson's disease or juvenile diabetes. Umbilical cord blood, which is rich in stem cells, can be banked, and used later in life, in case the person develops a disease, such as Parkinson's or Huntington's. Such research is at an early stage.

4

From Brain to Mind

The eyes provide the brain's window on the world. Vision provides most of our sensory input from the time of birth until death. More than 90% of all sensory input comes to us through our eyes. Sensations become perceptions that are the basis of thought, language, ideas, and concepts. Consciousness emerges from the activities in the brain, just as walking emerges from movement of the legs.

We can see the complexity of brain function by watching a baseball outfielder follow the course of a fly ball through the air, predicting where it will reach the ground, running to that spot, and catching the ball. The background against which the ball is moving is perceived as stationary. Wow!

The pupil acts as a pinhole through which photons of light pass through to the lens and reach the retina, where thousands of neurons fire and information passes across synapses in the rest of the brain. The chemistry of these synapses is the principal focus of molecular imaging.

In the retina, there are two kinds of light-sensing cells, rods and cones, that convert incoming photons into neuronal action potentials. When cones are activated by photons of light, pigment molecules change shape and produce electrical action potentials that pass information throughout the brain.

The pigments in the rods and cones respond to different wavelengths of light. The rods contain the pigment rhodopsin, or visual purple that responds to all incident photons, letting us see shades of black and white, even in dim light. The cones contain iodopsin. The cones are less sensitive to light than the rods, with the result that we can not see colors in the dark. Each rod has a single receptor, whereas there are three different receptors on cones. Each contains a different pigment that results in their absorbing light of different wavelengths: blue, green, and red. In the human eye, 7 million cones lead to the perception of high-resolution color images; the 120 million rods are extremely sensitive to white light.

Millions of neuronal fibers carry action potentials from the retina to the visual cortex and subcortical regions. If the eye is damaged at a time when the visual cortex is developing, there will be lasting impairment of vision. Depth perception is chiefly the result of differences in the input from each eye. One eye becomes dominant early in childhood.

Nobel Prize winners David Hubel and Torsten Weisel discovered how we perceive images. Neuronal genes encode the responses of different photoreceptors and send information to the primary visual area about the spatial orientation of lines, texture, retinal disparity, and color photons that reach the eye. The visual association cortex assembles, interprets, and stores these characteristics in short-term memory.

Hubel and Weisel measured the electrical activity of single neurons within the visual cortex of sleeping and waking cats, while the cats were looking at different light and dark patterns projected on a screen. Action potentials from different neurons in different parts of the eye respond to light or dark, the edges of objects, or motion. Some groups of neurons fire when the cat is looking at straight lines lying at different angles. Others respond to lines of different lengths. Others respond to different colors, and some respond to the direction of movement of lines. Neurons that respond to lines with a specific orientation are in the striate cortex (area IV) along horizontal lines and perpendicular columns. Perception of color completes the creating of emergent consciousness by the brain. Nowhere in the brain is there an image analogous to those obtained by a camera.

Sir Charles Sherrington, who was awarded the Nobel Prize in Medicine in 1932, wrote, "the eye sends into the cell-and-fiber forest of the brain throughout the waking day continual rhythmic streams of tiny, individually evanescent, electrical potentials. The large numbers of neurons throughout the brain that are activated by a scene do not resemble the way a scene would appear in a photograph. The neuronal distribution bears no obvious semblance to a spatial pattern. Even temporal relations resemble remotely the tiny two-dimensional upside-down picture of the outside world which the eyeball paints on the beginnings of its nerve fibers in the brain. When sunlight falls on an object, such as an apple, the fruit's skin absorbs most of the light, except for the red light waves, which are reflected off the surface of the apple and absorbed by the red cones of the retina. Photoreceptor cells are activated to transmit spatial information in color via countless axons and dendrites. Two genes in human beings encode color receptors. Defects in red and green color vision are X-linked" (Nathans et al., 1986). Both short and long-term memory involve many cortical streams of neurons with multiple connections.

Stephen Kosslyn used positron emission tomography (PET) and functional magnetic resonance imaging to map the neuronal networks related to vision. When a person views a scene, specific regions of the brain are activated. In a PET study with oxygen-15 labeled water, specific regions of the visual cortex became activated during the visual process. Visual imaging and visual perception involve the same regions of the brain. Looking at different small letters increased blood flow to small areas of the visual cortex corresponding to the fovea of the retina, whereas large letters activated neurons in the peripheral visual cortex.

Certain areas in the primary visual area (PVA) were more intensely activated when the subjects imagined a scene, compared with when they actually viewed the scene.

Perception begins with topographic localization in the occipital cortex of retinal representations, and progresses to visual memories in the temporal cortex. (Farah, 2000). Neuronal pathways connecting different parts of the visual cortex run in both directions.

Afferent nerves carry information from the eyes to the PVA and on to the visual association cortex, whereas efferent nerves send action potentials in the opposite direction. Stored visual memories in the visual associate cortex travel backward to the PVA, evoking a pattern of activity in the topographically mapped areas to produce a perceived image. Because this image originated from memory, it is a perceived mental picture, not a sensation.

Using ^{18}F-fluorodeoxyglucose (FDG)/PET, T. M.Bosley and colleagues examined the visual cortex of patients recovering from strokes that caused visual field defects (Ann. Neurol. 1987 21:444–450). All patients had striking impairment of glucose metabolism in the striate cortex. As the patients recovered, the area of decreased FDG accumulation in regions outside of the visual cortex became smaller. Those patients who had primary damage to the occipital lobe did not improve.

Thus, the brain is a "multilevel hierarchy of regions in which both serial and parallel processing occur simultaneously" (Corbetta et al., 2005). Changes in regional cerebral blood flow reflect changes in neuronal activity in adjacent cortical or subcortical regions of the brain. For example, randomly moving dots activated a region at the temporal-occipital-parietal junction.

4.1. Mental Illness

At 9:51 on the morning of October 2, 2006, Charles C. Roberts, a 32–year-old milk truck driver, walked into a one-room Amish School in Pennsylvania carrying a 9-mm handgun, 12-gauge shotgun, .30–06 bolt-action rifle, about 600 rounds of ammunition, cans of black powder, a stun gun, two knives, a change of clothes, a sexual lubricant, a truss board and a box containing a hammer, hacksaw, pliers, wire, screws, bolts, and tape. He barricaded the school doors and bound up the arms and legs of the terrified boys and girls huddled in the room. He released 15 male students, and a pregnant assistant teacher, but kept all 10 female students inside the room. As police broke in through the windows, he killed four of the girls and then shot himself. Six girls were taken to the hospital in critical condition. Most people expected that there would be violent, angry responses by the Amish community when they learned of the horrible event. They were surprised to learn that the responses focused on forgiveness, faith in God, and a determination to more forward. Lizzy Fisher, an Amish great grandparent of one of the slain girls, was asked whether the community felt angry about the killing. She answered, "Oh, no, no, definitely not. People don't feel that around here. We just don't." The Amish were consumed by sorrow, not anger (Urbana, 2002).

Weeks after the shooting, relatives of the victims, their neighbors, and volunteers built a new school a few hundred feet from where the shooting had occurred. The old school was torn down. "For the families and the community, it's a work of love and caring for the children, that becomes part of the healing process." The Amish viewed the building of a new school an act of love that brought the community together.

Why did the Amish community respond in this way? Did cultural factors and moral principles account for their emotional response? For hundreds of years, generations of Amish children have been taught not to lie, steal, or kill. It is possible that this was long enough for genes associated with altruism to provide a reproductive advantage. Or were their responses the result of cultural factors alone? Admiration of the ideas of peace, serenity, simplicity, veracity, and their suppression of the emotions of anger and revenge certainly affect their choice of marriage partners. Over hundreds of generations, genetic mutations may have enhanced the survival value of innate neuronal pathways and neurochemical processes related to altruism, while genes that encoded the emotions of extreme anger and violence became less prevalent. Today, such questions might be addressed by ethically designed research, by using molecular imaging of the brain.

Genes affecting emotions, such as hunger, sleeping, and sexual activity, were present in animals long before the evolution of humans. Genes evolved in humans that gave them the ability to think, understand, imagine, and speak. Innate aggressiveness and competitive behavior help preserve species, and have survived, at times resulting in violence, crime, and war.

No one has identified genes encoding morality. Nevertheless, character traits, such as conscientiousness and agreeableness, are found to be the same in identical twins separated at birth, and growing up in different environments. Some with antisocial personality disorder show signs of morality blindness as they grow up. They bully younger children, torture animals, lie, and are incapable of empathy or remorse, despite normal family surroundings. Some grow up to be criminals who try to talk elderly people out of their savings, rape women, or shoot convenience-store clerks lying on the floor during a robbery.

Psychological abnormalities at times result from head trauma that damages the frontal lobes. Some children who sustain severe injuries to their frontal lobes can grow up to be callous and irresponsible, despite having normal intelligence (Hanna and Antonio Damasio). They lie, steal, ignore punishment, endanger their own children, and cannot face up to even the simplest moral dilemmas, for example, deciding who should select which TV channel to watch. Damasio believes that this is evidence that a moral sense is partly innate.

Two months after the dreadful occurrence at the Amish school in Pennsylvania in 2006, day laborers gathered in the early morning in Iraq to wait for offers of temporary work. Shortly before 7 a.m., a small truck drove up, loaded with bags of wheat. The driver got out of the truck and shouted, "I need help to unload the truck!" When a crowd of men had gathered around, the driver set off a massive explosion, killing 70 and wounding 236 men, digging a crater 10 ft wide, and scattering wheat in all directions. The wounded were taken to the emergency room of a local hospital, short-handed because a large number of doctors and nurses had fled the country. Was the driver mentally ill? Was he searching for thrills and excitement, or acting according to his moral principles?

"Two things fill the mind with ever new and increasing admiration and awe, the oftener and more steadily we reflect on them," wrote Immanuel Kant, "the starry heavens above and the moral law within."

In his speech to Congress on March 18, 1917, after declaring war against Germany, President Woodrow Wilson said that "… wars are provoked and waged in the interest of dynasties or of little groups of ambitious men who were accustomed to use their fellow men as pawns and tools … The world must be made safe for democracy … We desire no conquest, no dominion."

Enthusiastic young men joined the armed forces in the United States after the bombing of Pearl Harbor on December 7, 1941, whereas some did all they could to escape the draft. All of their lives, they had been taught that killing is a sin. After a few months in battle, the soldiers began to view killing as a skill, even an art. Eighteen year-olds became adults, flying airplanes, commanding submarines, and tramping through unbelievably difficult terrains, surrounded by incessant noise and death. They developed close emotional ties with their colleagues, and they were willing to sacrifice their lives to save them.

U-boat warfare, and the bombing of Hamburg, Berlin, Dresden, and other European cities, Hiroshima and Nagasaki provide further evidence that extreme aggressiveness lies dormant even in people in the most advanced societies. Five million Germans died during War II, including 1.8 million civilians (MacDonough, 2007). The firebombing of Dresden and Hamburg killed hundreds of thousands of people, among them 75,000 children under 14 years of age.

Between 60 and 60 million war-related deaths occurred during World War II. The World Wars, and the Korean and Vietnam Wars, made the twentieth century the most bloody of all times (Ferguson, 2007). These wars witnessed great levels of atrociousness, barbarity, cruelties, and genocides.

Ferguson argues that extreme violence results from an explosive mix of "ethnic conflict, economic volatility, and empires in decline." Religious, racial, and linguistic minorities often suffer from economic depression, and resort to violence. Wars are brought about by widespread unemployment and poverty, aggravated by the destruction of crops, burning of villages, looting of towns, mass murders, barbarism, invasions and ever-changing national boundaries. Ferguson does not believe that aggressiveness is innate, but is the result of cultural flaws.

In November, 1945, Robert H. Jackson, leading prosecutor of the Nazi leaders at Nuremberg, said that those persons on trial "built up Adolph Hitler and vested in his psychopathic personality … the supreme issue of war or peace. They intoxicated him with power and adulation. They fed his hate and aroused his fears."

In January, 1971, Idi Amin Dada took power after a military coup in Uganda. His rule was characterized by human rights abuses, political repression, ethnic persecution, extra judicial killings, and the expulsion of Indians from Uganda. It is not known how many people were killed as a result of his regime but estimates range from 80,000 to 500,000.

Further evidence that leaders of nations can be mentally ill is provided by Saddam Hussein of Iraq, Osama bin Laden of Afghanistan, and President Ahmajinadad of Iran.

It is probable that terrorism is the result of innate aggressiveness brought to the surface by cultural and environmental forces. Pent-up anger begets murder and self-destruction. With weapons of mass destruction, terrorist gangs or demented psychopaths could destroy civilization, or even the human race.

Political efforts have been unable to solve the problems of killings, war, and destruction. The League of Nations, the United Nations, treaties outlawing war, and arms control agreements have all failed. Only strength and deterrence by individual nations brought peace after World War II and produced democracy in Germany, Japan, Italy, and Austria.

Today, the world remains plagued by violence. Tens of thousands of young men and some women have died in the process of killing others, while their colleagues applaud their martyrdom. With weapons of mass destruction, even a few people can inflict enormous harm on millions of people. Today, terrorism is a major threat to humanity. With molecular imaging, the minds and behavior of terrorists can become a major focus of study and research.

Humiliated terrorists often seek revenge against those they believe hurt them. In her book *What Terrorists Want: Understanding The Enemy, Containing the Threat* (Random House, 2006) Louise Richardson wrote, "The point of terrorism is not to defeat the enemy but to send a message." Justice, rather than freedom, is emphasized in the Koran. The goal is to end bureaucratic corruption, fight against social injustice and fulfill God's will.

According to Tariq Ramadan, Professor of Islamic Studies at Oxford (Ramadan, 2008), "The Holy Koran stands as the text of reference, the source and the essence of the message transmitted to humanity by the creator … It is the last of a lengthy series of revelations addressed to humans down through history … It is a light that responds to the quest for meaning … It is both the Voice and the Path … God is speaking to each [Muslim] in his own language, accessibly, to match his intelligence, his heart, his questions." A famous verse of the Holy Koran tells believers that slaying innocent persons is like slaying all of mankind unless it is done to punish villainy. Radical Muslims use this verse to justify aggression, murder and suicide.

When reading the Holy Koran, such as those verses that refer to war, one must reflect on the time and circumstances when these words were written. Only radical Muslims believe that injustices can only be faced by aggressive behavior, even war. "We are free and fully authorized to reform the injustices that lie at the heart of the order or disorder of all that is human" (Tariq Ramadan, a Swiss Muslim, who tells Muslims in Europe that they must establish a new "European Islam".)

Lee Harris, in his book *The Suicide of Reason* (Harris, 2007), believes that Islamic fanaticism today represents a struggle for cultural survival in the face of a rapidly changing world. The aim of Muslims is to preserve their culture and try to convert others all over the world. Islamic fanaticism is an attempt to expand the Islamic religion by violence if necessary.

Aggressiveness is viewed as a universal human trait, present long before human beings developed language and reasoning. The lives of early human beings, like other primates, were driven by fear. It was necessary to be selfish, aggressive, and combative to survive (Wade, 2007).

Thus, Islamic fanaticism should not be considered a mental illness. Harris writes, "The Muslims are, from an early age, indoctrinated into a shaming code that demands a fanatical rejection of anything that threatens to subvert the supremacy of Islam … they encourage their alpha boys to be tough, aggressive and ruthless."

The Enlightenment was an eighteenth century movement in Western philosophy that recognized the sad state of the human civilizations and the need for major reforms. Reason was to be the basis of authority.

Since then, there has been a reproductive advantage to those whose genes encode altruistic, rather than aggressive or violent behavior. Survival and reproduction are enhanced by societal rules, freedom of the individual, and secular governments. Genes that encode kindness and altruism supplement those encoding selfishness and violence.

Jonathan Haidt, a psychologist from the University of Virginia, proposes a happiness hypothesis, the idea that altruism evolved to enhance societal living. Genes encode loyalty to one's group and respect for authority. Morality evolves by natural selection in the face of human beings competing aggressively with one another. A principle evolved that "One should do no harm and do unto others as you would have them do unto you."

Ludwig von Mises (1881–1973), an Austrian economist, in his book *Human Action*, developed the idea that a person's every conscious action is intended to improve his or her personal satisfaction. Each person has his or her own idea of what personal satisfaction is. People act to remove sources of dissatisfaction and uneasiness. Because they are capable of logical thinking, they need not react instinctively to all stimuli, although dissatisfaction often drives people to act.

Michael Sherman, in his book *The Mind of The Market*, describes what he calls virtue economics. People are not only selfish, but are altruistic, cooperative and competitive, and peaceful. The balance between good and evil is heavily on the side of good. After societies began to form and people began to gather, altruism began to evolve because of its survival value. The innate golden rule increased reproductive activities. Working together and dividing tasks led people to treat others the way they themselves wished to be treated.

Gradually, government and religion codified proper behavior and advanced altruism. For every act of violence, there were 10,000 acts of kindness. Modern economies can function only because of people's innate virtuous nature. The motto Don't Be Evil has become a rule in commerce, when people make decisions in efforts to remove sources of dissatisfaction.

The belief that altruism has been the result of natural selection goes back to Charles Darwin, who said that people would risk their lives to save others. People are programmed by their genes to help their family. Altruistic genes are passed on from one generation to another because cooperation gives people a reproductive advantage.

William James believed that "Life is a challenge which everyone answers by their actions ... evil is something that we can help overthrow ... War gives angry self-confidence to millions of good people who have been taught to regard themselves as worthless." The capacity to be evil lies latent in every human being. Genes encoding good and evil behavior underlie the face of societal and cultural forces, which unfortunately can lead to aggressiveness, or even war. The Holocaust is evidence of the existence of man's inhumanity to man.

Good and bad persons differ in what it takes to activate the genes that express altruism or violence. Some people can become so angry that they would rather fight than accept what they view as intolerable. They can act like savages if sufficiently provoked.

Strong evidence that human beings are inclined toward evil is provided by fascism and other horrors of World War II. Even today, negative campaigning in elections appeals to the public's innate selfishness, where the candidates present the election as a conflict between good and evil.

"Is there anyone who will deny the existence of evil, or the important role of human fear and anger in its cause" (Jessiac Reyes, B.E. Journal of Economic Analysis and Policy).

Wise (2008) rejects the idea that people are really good at heart. She criticized the stage version of the *Diary of Anne Frank*. "European Jews before and during Hitler's reign may not have fully credited the human potential for evil, but surely the surviving remnant among them ought to be less willfully naive about human proclivities."

Many religions accept the concept of just wars. In his book *Moral Man and Immoral Society* (Scribner, 1934), Reinhart Niebuhr rejected the idea that evil can be accommodated. He believed that selfishness could only be controlled by self-discipline and love. Peace is a fragile, unsteady state, because tribes, movements, nations try to dominate other groups. The acceptance of force as justifiable has resulted in a nearly perpetual state of war in the twentieth century. Our only hope lies in a continual embrace of a biblically inspired faith in liberty, democracy, and in the spreading of these principles throughout the world (David Gelernter. Commentary, November, 2007).

Fouad Ajami has said that radical Islamists "have declared nothing less than an unrelenting war against the American presence in the Arab-Islamic world ... The region they contest is the Arab and Persian heartland of the Islamic world and cannot be ceded to them, for its obvious importance to the global economy."

He views Islam as a faith and civilization locked in a brutal struggle with Christians, Jews and secular humanists of the West. Radical Islamists believe that the millions of Muslims who are backsliders have consumed the worst of the secular Western world. They loathe the secularism and want to totally extirpate Western culture from their societies.

The 12,000 Madrasses in Pakistan teach thousands of children to hate America and Israel. The army is divided between the Western oriented, who have often attended schools in the United States, and hard-line Islamists. General Hamid Gul hates America with a passion, and is an Islamic extremist and strategic adviser to the six political religious parties that govern two of Pakistan's four provinces.

There are 12–14 million Islamic extremists throughout the world today. Iraq has been the central focus of the war on terror, but today Pakistan and Iran are in danger of succumbing to Muslim extremism.

A great fear is that Islamist extremists within the Pakistani army will take control of nuclear weapons. Hundreds of years of war followed the rise of Islam in the seventh century. Today, most Islamists reject the ideas of fundamentalists. Most want a peaceful coexistence with the West.

"The faith of Islam teaches moral responsibility that enables men and women, and forbids the shedding of innocent blood. This is a clash of political visions." (President George Bush). He believes that the West is in a battle for survival in the struggle with radical Islamists. Facing a new dark age, Muslims and non-Muslims need to come together on equal terms. A thesis of this book is that perhaps a better understanding of the chemistry of the brain and its relationship to behavior can enhance social, cultural and political forces in the interest of all humankind.

4.2. Reason for Hope

Molecular imaging can be used to relate mental illness, including violence and excessive aggressiveness, to specific abnormalities of molecular processes in different parts of the brain. PET, single photon emission computed tomography (SPECT), or optical imaging measure regional molecular processes.

Molecular phenotypes are being used more and more in diagnosis, prognosis, and treatment planning, as well as in the assessment of the results of treatment, providing long-term follow up. If the treatment is going to be chemical, perhaps the diagnosis should be chemical, i.e., the finding of deficient or excessive molecular processes in different parts of the brain. Appropriate medication, based on molecular diagnosis, can often restore abnormal regional or global chemical processes to healthy levels.

When aggressive and violent behavior becomes extreme, this can be viewed as disease. For example, drug addiction is now considered a "disease"; syphilis is no longer called wanton behavior but is identified as a sexually transmitted disease or a sexually transmitted infection.

Molecular dimensions can be added to the definition of mental diseases. Using Bayes's Theorem, we can calculate the probability of a specific biochemical manifestation being related to specific manifestations of a mental disease. For example, one can calculate the probability of a person's aggressiveness being related to epinephrine-related brain processes.

The idea that psychiatric diseases are caused by biological abnormalities goes back to Emil Kraepelin, a German psychiatrist, the founder of scientific psychiatry and psychopharmacology. He defined diseases by grouping patients with similar symptoms, signs, laboratory findings, or abnormalities detected at autopsy. Putting a patient in the correct group was called making the diagnosis, and it provided a basis for prognosis and treatment.

Medical diagnosis is based on probabilities. The physician forms a priori hypotheses about what is wrong with the patient. These are called working diagnoses, often based on psychological manifestations of disease. According to a molecular theory of mental disease, the results of PET/computerized tomography (CT), SPECT/CT, and other imaging technologies can help determine the a posteriori probability that the patient's psychological disturbances are related to molecular abnormalities in the brain. All evidence related to the patient's problem can affect the plan for treatment.

Bayes's Theorem requires that separate bits of information about the patient, including all clinical and laboratory information, be relatively independent of each other. Molecular imaging results provide important, quantifiable data. The physician integrates the findings with archival data concerning all possible mental diseases, and then makes a molecular diagnosis.

A problem in the diagnostic process is the result of the considerable variability in the manifestations of disease among different patients. For example, patients diagnosed as having diabetes insipidus have varying degrees of deficiency of antidiuretic hormone. The levels of osmolality in their urine varies greatly depending on the degree of deficiency of antidiuretic hormone and the amount of solutes in the urine.

5

The Early Days

Claude Bernard (1813–1878) dedicated his life "to the search for an understanding, in terms of physics and chemistry, of those processes by which we live, by which we become ill, by which we are healed, and by which we die." His research transformed medical science from its anatomical to a new foundation in physiology.

In 1878, the Cambridge physiologist, John Newport Langley, proposed that drugs act by binding to molecular receptors within the body (J. Physiol. [Lond.] 33:374–413). The inhibiting effect of atropine on the effect of pilocarpine was attributed to "some substance in the nerve endings or gland cells with which both atropine and pilocarpine are capable of forming compounds." He called them receptive substances. We now call them receptors.

In 1890, William James wrote, "Chemical action must of course accompany brain activity, but little is known of its exact nature." What would James think about today's molecular imaging of the brain?

In 1905, Paul Ehrlich was the first to use the word *receptor* (Ehrlich, 1962). His studies of arsenicals in the treatment of syphilis led him to postulate that the amplification of the effect of arsenic depended on receptors. He called salvarsan a magic bullet, because it was effective even when given in exceedingly small amounts. Recognition sites bind drugs, amplifying their effects, and regulating major processes throughout the body.

Seymour Kety was the first to measure the blood flow to the brain of a human being, applying the Fick principle after injecting nitrous oxide (N_2O), a diffusible gas, into a cerebral artery, and then measuring the difference in tracer concentrations in the arterial blood going to the brain and in venous blood coming from the brain.

Arterial blood was sampled from the femoral artery to reflect cerebral arterial blood. Venous blood was sampled from the superior bulb of the internal jugular vein. After the subject had breathed nitrous oxide for 10 minutes, the concentrations in the brain and cerebral venous blood were close enough to equilibrium to allow calculation of brain N_2O concentration from the measured cerebral venous concentration at that time and the relative solubility (i.e., partition coefficient) of N_2O in brain and blood.

In 1913, George Hevesy introduced the use of radioactive indicators when he used radioactive lead to track the movement of lead from the soil into plants and then back to the soil. He later extended these studies to living animals. In 1935, he showed how elements and molecules are taken up and released continually by living cells, a process called the dynamic state of body constituents.

The first cyclotron was invented by Ernest Lawrence and his graduate student M. Stanley Livingston in the 1930s, a decade before the invention of the nuclear reactor by Enrico Fermi. In 1931, Lawrence succeeded in accelerating positively charged hydrogen ions to bombard target atoms at high energies to overcome the nuclear charge. He and his colleagues produced one larger cyclotron after another, eventually building a 60-in. cyclotron, called the medical cyclotron.

On February 10, 1934, Frederick Joliot and Irene Curie published their discovery that practically every chemical element could be made radioactive. After hearing this news, Lawrence's group focused on the positron-emitting radionuclides: Nitrogen-13, carbon-11, and oxygen-15. These radionuclides, plus fluorine-18, which can serve as an analog of hydrogen, provide the foundation of molecular imaging.

On Christmas Eve of 1936, after successfully treating mice with leukemia, Ernest Lawrence's brother, John Lawrence, carried out the first radiopharmaceutical therapy when he administered a radiophosphorus solution to a 28-year-old woman with leukemia. Subsequently, he used radiophosphorus to treat patients with polycythemia vera.

Another major figure in the history of molecular imaging is William G. Myers, whose Ph.D. thesis at Ohio State University in 1941 was entitled, Applications of the Cyclotron and Its Products in Biomedicine. Despite his best efforts, he was not able to persuade the chairman of radiology to install a cyclotron at Ohio State.

At the 14th annual meeting of the Society of Nuclear Medicine in Seattle, WA, in 1967, William G. Myers reported that, "Intravenous injection of C-11 insoluble carbonates ... provide particles for scanning lungs ...". He pointed out that C-11 should achieve major emphasis in nuclear medicine because 18% of the human body is carbon.

Fermi and colleagues at the University of Chicago put cyclotrons on a back burner in biomedical research by inventing the nuclear reactor during World War II, which made possible the production of large amounts of carbon-14, tritium, phosphorus-32, and other radionuclides. Carbon-14 became the foundation of the field of biochemistry.

In June, 1946, the American Atomic Energy Commission announced in the journal Science that they would provide cyclotron- and nuclear reactor-produced radionuclides for scientific research to qualified persons throughout the world. The first shipment of a C-14 tracer was to the Bernard Skin and Cancer Hospital in St. Louis.

The longer half-lives of these reactor-produced tracers facilitated their availability and decreased their cost compared to cyclotron-produced radionuclides. Carbon-14 and tritium emitted only beta particles with a very short range in tissue, so they could not be used to examine regional biochemistry with radiation detectors pointed at the human body. Three decades were to elapse before physicians and scientists returned to the cyclotron as an important, indeed essential, tool in molecular imaging.

In an article in NUCLEONICS in 1966, Michel TerPogossian (Figs. 5.1–5.3) and I wrote, "The most important radioactive tracer in biological research is reactor-produced carbon-14, but it has never been widely used in nuclear medicine ... The use of these nuclides (Carbon-11, Nitrogen-13, Oxygen-15 and Fluorine-18) in biomedicine justifies the additional effort to prepare them locally at the laboratory or hospital that plans to use them."

The term atomic medicine was used for decades after the end of World War II to describe the new field of medicine based on the use of radioactive tracers to examine the dynamic state of body constituents.

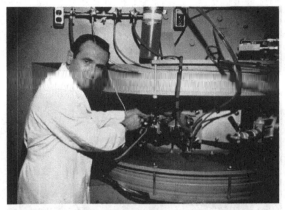

Fig. 5.1 Michel TerPogossian wi thhi sc yclotrona t WashingtonU niversity.

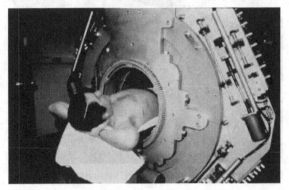

Fig. 5.2 Henry Wagner having a PET scan at Washington University in 1975.

In 1948, Robert Reid Newell of Stanford University proposed that the name be changed to nuclear medicine. Radiologists, including David Kuhl and Merrill Bender, played an important role in the early days of nuclear medicine, so there was an emphasis on body structure. Radioisotope scanning was defined as visualizing organs and lesions not visible on conventional x-rays.

In the early 1950s, Benedict Cassen (Figs. 5.4–5.6) at the University of California developed the first rectilinear scanner based on the use of radiotracers produced by a nuclear reactor. Later, Kuhl and Bender invented a method called photorecording of images on x-ray film.

Radioisotope scanning was used in the early days to detect filling defects in organs, such as the liver and spleen. Radioactive sodium iodide was used to image the thyroid. Radioactive particles were used to image the reticuloendothelial system. The dye, radio-iodinated Rose Bengal, was used to image the liver.

In the 1620s, the French philosopher, Rene Descartes. coined the word molecule. The existence of molecules was not accepted until the early nineteenth century when John Dalton proposed an Atomic theory in 1803, which stated that all matter was composed of small indivisible particles called atoms. Avogadro's law extended Descartes' law of

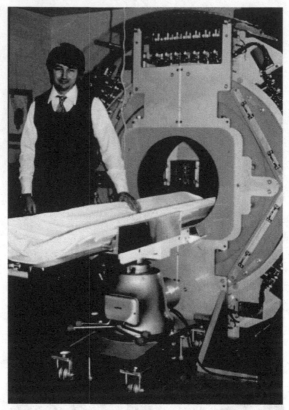

Fig. 5.3 Michael Phelps, early inventor of a PET scanner at Washington University.

multiple proportions in 1811. He hypothesized that equal volumes of ideal gases, at the same temperature and pressure, contain the same number of molecules.

Billions of different molecular processes are going on in the body at all times. These processes are the focus of molecular imaging.

The researchers extended the Kety–Schmidt method to the use of chemically inert radioactive tracers, krypton-79 and xenon-133, because it was easier to measure their concentrations in blood. Later, oxygen-15 and xenon-133 were used to measure regional blood flow to the brain. Oxygen-15 made it possible to measure the rate of metabolism of oxygen to carbon dioxide. Oxygen-15 is produced in a cyclotron, its use complicated by its short (2.5 min) half-life.

In 1940, the invention of the nuclear reactor was a major advance. It was able to produce the longer lived isotope, carbon-14, that led to the development of modern biochemistry. The reactor led to a decreased use of cyclotron-produced radionuclides, even though the cyclotron had the important advantage in producing the positron-emitting tracers, oxygen-15, carbon-11, fluorine-18, and nitrogen-13. Bombardment with subatomic particles, including, protons, in a cyclotron produced these radionuclides.

In 1956, Hugh Stoddart of the Atomic Instrument Company, in collaboration with Brownell and Sweet at Massachusetts General Hospital, developed the first commercial

Fig. 5.4 Benedict Cassen, inventor of the rectilinear scanner in the early 1950s.

Fig. 5.5 Rectilinear scan of a patient with a stroke. The tracer was technetium-99m albumin.

brain scanner for imaging positron emitting radiotracers. Two 180° opposed radiation detectors were in a yoke that performed a back-and-forth movement over the patient's head. Coupled to the yoke was a solenoid that would put marks on paper corresponding to the recorded disintegration of the tracer, arsenic-74.

In the 1970s, there were 10 positron emission tomography (PET) scanners and cyclotrons in the United States, producing radiotracers to obtain images of regional blood flow and chemistry in the human brain. In 1971, in a lecture at the ceremony dedicating the new cyclotron at the University of California in Los Angeles, I said, "the day may come when cyclotrons will be common, at least in larger teaching hospitals. It is not too difficult to envisage a time when every hospital will have access to short-lived nuclides for tomographic studies routinely. It will be some years before this comes to pass, of course."

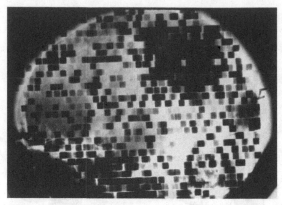

Fig. 5.6 A fused image of a brain tumor obtained with a rectilinear scanner in the early 1960s. The scan was performed with I-131 albumin and superimposed on a conventional radiograph of the skull.

Fig. 5.7 First PET scanner, designed by Yamamoto of the Brookhaven National Laboratory in the 1970s.

Today, nearly four decades later, carbon-11 and fluorine-18 are widely used to examine regional biochemical processes in the living human brain. Either PET or single photon emission computed tomography (SPECT) can reveal the phenotypic expression of genetic abnormalities, often expressed as from genotype to phenotype.

Many companies throughout the world produce PET and SPECT instruments, today almost always with a computerized tomography or magnetic resonance imaging capability in the same gantry.

Hamamatsu Photonics K.K. is a Japanese company that for decades has made enormous contributions to the growth and development of molecular imaging. Teruo Hiruma, its president and one of its founders, followed in the footsteps of Professor Kenjiro

Takayanagi, who introduced television in Japan in 1926. Mr. Hiruma calls the twenty-first century the Century of Light. "Photons reveal the chemical melodies sung by the orchestra of the trillions of cells within us."

In 1987, a group of Japanese, Americans, and others founded The Research Foundation for Optical Science and Technology in Hamamatsu City, Japan. Its goal is "to promote scientific collaborative research on the brain mechanisms of violence and destructive behavior in human beings and to make such research a major focus of collaboration between scientists and engineers of Japan, the Soviet Union and the United States." At the time, a Cold War was going on between the Soviet Union and the United States. The focus of the Foundation is on science, not politics. The main foci of the meetings held every 18 months has been brain imaging and genetics.

The idea of the foundation began during a visit by Mr. Teruo Hiruma to Johns Hopkins in 1986. He was interested in an idea I had presented at a nuclear medicine meeting in Korea the year before. I had described the potential for brain imaging in trying to understand the biochemical correlates of aggressive behavior. Aggressiveness, violence and wars were prevalent throughout the world, and Mr. Hiruma asked whether the increasing scientific understanding of brain chemistry might help solve problems, such as the Cold War, avert a catastrophy, and create a more peaceful world.

A goal of the Foundation was to promote scholarly and scientific exchanges and joint research projects among physicians and scientists from the Soviet Union, Japan, the United States and other countries. The founding document stated, "A testable hypothesis is whether mind/brain science can help explain violent and overly aggressive behavior. Violence can be viewed not only as socioeconomic and political phenomena, but also as a form of mental illness, subject to investigation by means of biochemical, electrical and psychological methods."

Starting in the 1950s, radioisotope scanning created images of the distribution within the brain of injected radioactive tracers. Hertz and his associates in Boston, and Hamilton and associates in Los Angeles were the first to diagnose increased or decreased thyroid function by measuring the uptake of radioactive iodine-128 by the thyroid, and radioisotope scanning of the thyroid made it possible to tell whether thyroid nodules were functional and produced thyroxine. This information helped in determining whether they were malignant, and thus should be removed.

Since then, radioactive tracer elements have been incorporated into innumerable metabolically active molecules, which are then injected intravenously and become concentrated in regions of interest. PETracers emit 511-keV photons imaged by a circle of radiation detectors surrounding the patient's head. A computer constructs images of the distribution of the tracer within the brain at various times after injection of the tracer.

Mathematical analysis of a series of images obtained over time yield what are called functional images. Time adds the important fourth dimension to the three-dimensional spatial distribution of the tracer. The rate of accumulation of the tracer in different regions of the brain, or some other mathematical analysis, is calculated for each picture element (pixel) in the images.

In SPECT imaging, single photons are emitted from the radionuclide in all directions. Spatial location of the emissions is obtained by means of lead collimators. As in PET, the choice of the radioactive tracer for a study depends on the molecular process being examined.

6

A Molecular Theory of Mental Disease

The technologies positron emission tomography (PET) and single photon emission computed tomography (SPECT) made it possible to move the orientation of medical science from organs, tissues, and circulating blood to cells and molecules. This brought about a new view of disease. The germ theory of disease views disease as a struggle between a human host and offending bacterial organisms, many of which are normal flora that become infectious. Anton van Leeuwenhoek's invention of the microscope in the seventeenth century made it possible for the first time to see the bacteria themselves. PET and SPECT let us image the molecules within the body.

In 1876, Robert Koch proposed four postulates that needed to be met to establish that a specific microorganism was the cause of a disease: (1) the organism had to be present in all patients with the disease; (2) the organism had to be isolated from the patient; (3) when the organism is injected into a susceptible host, it must cause the same disease as that in the original host; and (4) the organism must be isolated a second time from the infected host.

The time has come to create a molecular theory of mental disease. Instead of putting patients into diagnostic pigeon holes, such as depression, Parkinson's disease, or schizophrenia, that are poorly defined, heterogeneous, and nonspecific, we can use molecular imaging to relate a patient's problems to regional brain chemistry.

Molecular imaging as the foundation of diagnosis is not new. Half a century ago, imaging the rate of accumulation of radioactive sodium iodide by the thyroid was the first example of molecular imaging, although it was called radioisotope scanning.

Whenever a molecular process is measured in some part of the human body, in theory, two diseases are possible—one disease in which the process is abnormally fast, and the other disease in which it is abnormally slow. In the thyroid, the diseases are called hyper and hypo-thyroidism.

The regional chemistry of molecules, such as epinephrine, dopamine (DA), or serotonin (5-hydroxytryptoamine, 5-HT), can now be examined in the brain of patients with mental illness. One day, a patient's diagnosis will be his name, characterized by a personal database containing all the patient's manifestations of health or disease.

As in infectious diseases, host factors are extremely important. Just as billions of normal flora make up 2–3% of the bacteria in an adult human body, so also are there billions of molecular processes in the normal brain. In deficient or excessive amounts, these molecules can cause disease, just as a few miscreant bacteria can produce an infectious disease.

Little is known about the role that the normal bacterial flora, fungi, and other microbes play in human health and disease. The National Institutes of Health (NIH) in Bethesda, MD, has just launched an official launch of the Human Microbiome Project, to determine the genomes of all microorganisms present in or on the human body. We need the same approach to examine the roles of thousands of types of molecules in the human brain. We need a molecular theory of mental disease, based on four postulates: (1) An external or internal threat brings about changes in the release of specific hormones or neurotransmitters. (2) The molecule(s) produce an emotional response. (3) The emotional response is accompanied by thoughts. (4) The emotions and thoughts bring about actions. Molecular diseases are characterized by emotions, thoughts, and actions. The fight-or-flight response to external threats, when extreme, is an example of a molecular disease.

In the 1950s, Joel Elkes and colleagues studied the effects of drugs, including amphetamines and lysergic acid diamide (LSD), on the electrical activity of the brain, and related their findings to behavior. They wrote, "Rather than thinking in unitary terms, it may, at this stage, be advisable to think in terms of the possible selection by chemical evolution of small families of closely related compounds, which by mutual interplay would govern the phenomena of excitation and inhibition in the central nervous system." Biochemical processes activate and inhibit the activity throughout neuronal networks in regions of the brain involved in visual, verbal, motor, and emotional functions.

The first molecules identified as transmitters of information within both the peripheral nervous system (PNS) and brain (CNS) were acetylcholine (ACh), norepinephrine (NE), and 5-HT. ACh is the neurotransmitter in the autonomic nervous system. Profound biological effects can result from even slight changes of molecular configuration.

Hundreds of neurotransmitters have since been identified. Some are "small-molecule neurotransmitters," including ACh, monoamines (NE, DA, and 5-HT), amino acids (primarily glutamic acid, GABA, aspartic acid, and glycine), purines, (adenosine, ATP, GTP, and their derivatives), fatty acids (endogenous cannabinoids), and peptides (vasopressin, somatostatin, and neurotensin). Hormones, such as insulin or thyroxine, have distant effects on the cells of the body, as well as local effects. Single ions, such as zinc, which is secreted into synapses, also are neurotransmitters.

Neurotransmitters are constantly being synthesized and stored in vesicles at the end of presynaptic axons. The release of neurotransmitters into the synapse from vesicles in the presynaptic neurons results from the opening of calcium channels. Calcium fuses with the presynaptic neuronal vesicles to bring about the release of the neurotransmitters into the synapse where they bind to receptors on postsynaptic neurons.

Neurotransmitters include catecholamines (NE), biogenic amines (5-HT and ACh), polypeptides/proteins (endorphins), and amino acids/amino acid analogs (glutamate and GABA). These molecules bring about control of the excitation and inhibition of neuronal activity. DA has an inhibitory effect, by reducing the number of postsynaptic action potentials. Other neurotransmitters have an excitatory effect.

6.1. Dopamine

In the 1950s, pharmacologist Arvid Carllson left Uppsala, Sweden, to come to the Laboratory of Chemical Pharmacology of the NIH. This decision was the best career choice he ever made. Steve Brodie at the NIH had begun to work with a new instrument, called a spectrophotofluorimeter, invented at the NIH by Sydney Udenfriend, which made it possible for the first time to measure very low levels of drugs, metabolites, and compounds, such as 5-HT, in body fluids and tissues, including the brain. In 1957, Arvid Carlsson (Fig. 6.1) used this device to discover that the monoamine DA was a neurotransmitter, a discovery for which he and Paul Greengard were subsequently awarded the Nobel Prize.

Carlsson discovered that DA is present in large amounts in the brain, and disappears when a person is given the drug reserpine. Administration of the precursor of DA, L-dihydroxyphenylalanine (L-dopa), brings about the reappearance of DA in the brain. Carlsson said, "I would put the discovery of dopamine and its role for normal brain functions as a winner, no doubt about it. A good second place would go to the research into the role of dopamine in mental disorders, such as schizophrenia, followed by the third finding: the mode of action of antipsychotic drugs."

The neuronal cell bodies of the dopaminergic neurons are in the pars compacta of the substantia nigra from where axons extend to the cerebral cortex and limbic regions. Dopaminergic neurons are also in the ventral tegmentum of the midbrain.

DA is involved in movement disorders, as well as in schizophrenia and substance abuse. The neurotransmitter is synthesized from a precursor amino acid tyrosine. L-tyrosine is hydroxylated to L-dopa by the action of the enzyme, tyrosine hydroxylase. L-dopa is then converted to DA by the L-aromatic amino acid enzyme decarboxylase.

DA is not only a neurotransmitter but also a precursor of NE and epinephrine. It is involved in movement, feelings of pleasure, and depression, as well as in schizophrenia, and attention deficit hyperactivity disorder.

Radiotracers have been developed that bind to pre- or postsynaptic receptors or reuptake transporters, or act as substrates for the enzymes involved in the synthesis or inactivation of neurotransmitters, such as DA.

To develop new drugs to affect mental activities, it is often is necessary to modify their molecular structure so that they can cross the blood-brain barrier. Metabolism and washout from the brain are also important.

Fig. 6. 1 Arvid Carlsson, who discovered that dopamine is a neurotransmitter.

In 1976, Solomon Snyder and colleagues at Johns Hopkins showed that the effectiveness of neuroleptic drugs, such as spiperone and haloperidol, was directly related to their affinity for binding to D_2 DA receptors. Seeman and others at the University of Toronto also reported this finding. These results raised the question of whether schizophrenic symptoms might be the result of excessive DA release or hypersensitivity of DA receptors—the DA hypothesis of schizophrenia.

When amphetamines are taken to suppress appetite, they often produce a psychosis indistinguishable from schizophrenia. D-Amphetamine (dextroamphetamine) and L-amphetamine, or a racemic mixture of the two isomers, binds to the DA and other monoamine transporters that increase synaptic levels of the DA, NE (noradrenaline), and 5-HT. D-Amphetamine acts primarily on the dopaminergic systems, whereas L-amphetamine acts on the norepinephrinergic neurons. The reinforcing and behavioral-stimulation effects of amphetamine are possibly the result of the enhanced dopaminergic activity in the mesolimbic system.

Snyder has said: "… many amphetamine users admitted to hospitals have been diagnosed as paranoid schizophrenic until the history of drug use was uncovered days or weeks later." Kuhar and associates at Johns Hopkins showed that DA receptors in the basal ganglia could be visualized by autoradiography after administration of drugs that were bound by DA receptors. This encouraged us to submit a proposal to the NIH to try to image DA receptors in living humans by PET.

Snyder and colleagues at Johns Hopkins had used carbon C-14 labeled drugs to prove the existence of neuroreceptors in the brain of rodents. Our goal was to label antipsychotic drugs with C-11, a positron-emitting radiotracer, that would make it possible to image their distribution in the brain with PET.

To do so, we needed to obtain a cyclotron and PET scanner and to expand our chemical synthesis laboratory. The NIH provided the funds needed to purchase the cyclotron, and the Johns Hopkins Hospital provided the funds to purchase a PET scanner. We made C-11 in a Swedish cyclotron that we purchased in 1978.

Dr. Bengt Langstrom of the University of Uppsala developed a method to label methyl spiperone, a neuroleptic drug, with C-11, and taught Hopkins chemists, led by Bob Dannals, how to carry out the synthesis during his visit to the Nuclear Medicine Division at Johns Hopkins. The manufacturer of the cyclotron, Scandatronix, sponsored Bengt's visit to Johns Hopkins.

The goal of our research program in the Nuclear Medicine Division of Johns Hopkins in the late 1970s was to study how brain chemistry is related to mental activity and behavior. Consciousness is analogous to the story in a book, where molecular processes are sentences, and molecules are the words. PET and SPECT let us read the book.

Our administrative leaders at Johns Hopkins were skeptical. On February 8, 1979, I received a letter from Dr. Bob Heyssell, Director of Johns Hopkins Hospital (before coming to Johns Hopkins, he had been head of nuclear medicine at Vanderbilt University).

He wrote, "I happened to see the article in *The Sun* in which you are quoted extensively concerning expansion of your brain study program for thought imaging … Don't we have enough problems as it is without our faculty and others helping to make us look like idiots or incompetents or worse?"

He had read an article written by Jon Franklin, a Pulitzer Prize-winning reporter from *The Baltimore Sun*, which said: "Johns Hopkins scientists say they will install a scanner capable of watching the chemical process of thought as it flickers through the brain ... It is designed to help map the circuitry of thought and allow scientists to understand how information is processed by the brain ... Once brain specialists understand how energy flows through the brain, they should be able to pinpoint the problem in patients with certain diseases, including schizophrenia, senility, Huntington's disease, Parkinson's disease, manic-depressive psychosis, drug addiction and manganese poisoning ... "

"Dr. Henry Wagner, the scientist in charge of the project, expects the device, a positron emission tomography (PET) scanner, to pinpoint chemical malfunctions in the brain and, for the first time, to allow scientists to tinker directly and rationally with the chemical engines of thought ... If we find something that runs too fast, we can treat it by slowing it down. If it's too slow, we can speed it up."

In early studies with C-11 N-methyl spiperone (C-11 NMSP), we found a striking decrease in DA receptor availability in the basal ganglia with advancing age, correlated with decreasing motor performance and emotionality. Starting at age 45, DA in the basal ganglia falls about 6–13% per decade. In patients with Parkinson's disease, it is below 30% of normal levels.

Since these first studies in 1983, many new radioactive tracers have been developed to study brain chemistry. They are used in the development of new drugs that affect mental activity. To visualize brain structure together with brain chemistry, we now fuse computed tomography (CT) and magnetic resonance imaging (MRI) images with PET and SPECT images. CT and MRI provide exquisite structural detail, whereas PET and SPECT make it possible to measure molecular processes within well-defined structures.

In our research, we most often label drugs with C-11 and study their pharmacology in baboons and then normal humans, before proceeding to study patients with mental illnesses, such as schizophrenia.

The process of methylation with C-11 methyl iodine eventually became widely used for labeling many drugs. To produce C-11, because of its 20-minute half-life, it is necessary to have a cyclotron close to where the chemical syntheses and imaging procedures are carried out.

Up until our obtaining a cyclotron, we developed single photon-emitting radiotracers to image the heart, spleen, liver, kidneys, and other organs. Our NIH-proposal was entitled *Short Lived Radionuclides in Biomedical Research*.

In the 1960s, we proposed to the NIH that we reactivate a 60-inch cyclotron built in 1939 by Merle Tuve at the Carnegie Institute of Washington. It was identical to the 60-inch cyclotron built by Ernest Lawrence in Berkeley, CA.

Our scientific proposal was approved for funding by the NIH. We also had requested funds to construct a new building to house the cyclotron. We did not get NIH approval to fund the building, possibly because 20 new medical schools were then being established in the United States, and NIH funds were needed for these new buildings. There was no existing space on the Johns Hopkins medical campus; so, our cyclotron project was postponed until the 1970s.

During his visit to Johns Hopkins in 1965 for the NIH site visit of our cyclotron proposal, Alfred Wolf, a chemist in charge of the cyclotron at the Brookhaven National Laboratory (BNL), saw the potential value of cyclotron-produced radionuclides in biomedical research. He and his colleagues, including Joanna Fowler, subsequently developed a large number of radioactive tracers, including fluorine-18 deoxyglucose (FDG), the tracer that was to propel PET into the mainstream of medicine.

   John Totter, head of the medical programs of the Atomic Energy Commission (AEC), was also one of the NIH site visitors. He, too, was convinced and had the AEC fund three hospital cyclotrons in the United States, run by physicists Michel TerPogossian (at Washington University in St. Louis in 1965), Gordon Brownell (at the Massachusetts General Hospital in Boston in 1967), and John Laughlin (at Memorial Hospital Sloan Kettering in New York in 1967). The cyclotron then began to move back into the forefront of nuclear medicine. By 1991, the cover of the *Journal of Nuclear Medicine* and an editorial proclaimed, "Clinical PET: Its Time Has Come."

   Molecular imaging requires both the proper radioactive tracer and an instrument for imaging the tracer distribution within the living human body. Kuhl and Edwards (1963) were the first to describe tomography in nuclear medicine. Their invention preceded that of Houndsfield using x-ray transmission tomography (Figs. 6.2 and 6.3).

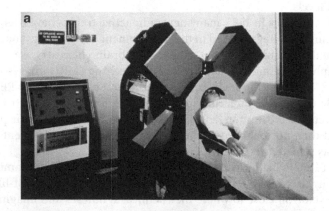

Fig. 6.2 **(a)** First tomographic radioisotope scanner, invented by David Kuhl at the University of Pennsylvania in the 1960s. The patient had a stroke, and the tracer was technetium-99m albumin. **(b)** Dr. David Kuhl and Dr. R. Q. Edwards. **(c)** Schematic drawing of the first tomographic scanner.

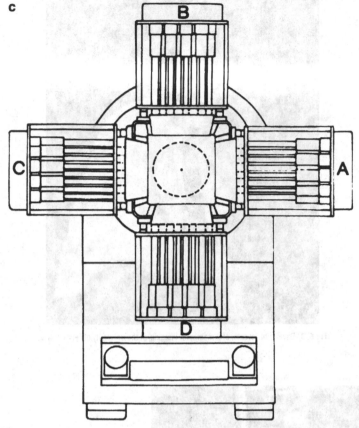

Fig. 6.2 (continued)

Kuhl's group carried out the first imaging of glucose utilization in the human brain with FDG, an analog of glucose. They found regional abnormalities in the use of brain glucose before there were demonstrable structural changes in patients with stroke, senile dementia of the Alzheimer type, and Huntington's disease.

In 1983, after the first imaging of a neuroreceptor (Figs. 6.4 and 6.5), the DA receptor, in the brain of a living human at Johns Hopkins, research the very successful moved beyond FDG studies.

Over the decades, when molecular imaging was growing, striking advances also were being made in molecular biology and genetics. Molecular biologists were isolating and characterizing the molecular components of living cells, including DNA, the repository of genetic information; RNA, the mechanism for duplicating DNA; and the synthesis of proteins and peptides.

Nobel laureate Sydney Brenner, said, "Modern biology has become so genocentric that we have forgotten that the real units of function and structure in an organism are cells and not genes …"

Fig.6. 3 David Kuhl, inventor of tomographic imaging, at the University of Pennsylvania.

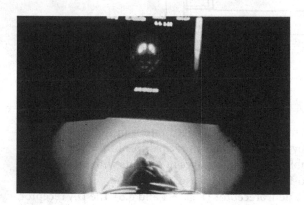

Fig. 6.4 First imaging of a neuroreceptor in a living human being. The tracer was C-11 N-methyl spiperone and the receptor was the D_2D Ar eceptor.

After the basic principles of information transfer from genes to proteins had been established with the identification of messenger RNA, the discovery of the mechanism of protein synthesis and the structure of genetic code, it was natural for some of us to ask whether the lessons learnt in molecular biology could be applied to the genetics of more complex phenotypes. Molecular imaging provides the bridge extending from genotype to phenotype.

Brenner advises students, "All you have to do is to find a gene and have it sequenced and then make some protein using the gene and get someone to determine its amino

Fig. 6.5 First imaging of opiate receptors. Left to right. Solomon Snyder, James Frost, and Henry Wagner (in scanner).

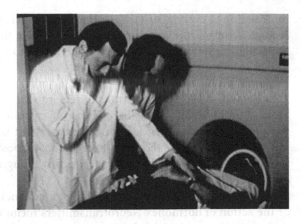

acid sequence." The next step is to determine the role that the proteins and other molecules play in the process of living. Molecular imaging makes this possible.

Recording electrical activity with electrodes surrounding the head made it possible to try to relate the mind to the brain. Slow electrical brain waves, called theta waves, were observed during states of low mental activity. At high degrees of arousal, faster beta waves were recorded. Alpha waves were recorded at medium levels of arousal. The greater the state of arousal, the greater the number of alpha and beta waves.

The electroencephalogram (EEG) reflects the billions of action potentials that travel along the neuronal dendrites and axons. Dendrites carried information toward neuronal cell bodies; axons carry information away from cell nuclei. The action potentials are created by sodium and potassium ion gradients across the cylindrical neuronal membranes, where there is more potassium on the inside than on the outside of the dendrites and axons.

In the human brain, there are at least 100 trillion synapses that connect 15–100 billion neurons. Each neuron is connected to thousand of others via synapses, where information traveling among neurons is modified by neurotransmitters.

Chloride ions inside dendrites and axons result in a negative transmembrane charge when the neuron is at rest. Action potentials that pass along nerve fibers are the result of successive opening of "gates" that let sodium flow from the outside to the inside of the nerve fibers. As the electrical signal passes along the fiber, potassium ions return from the outside to the inside of the neuronal fiber. Waves of transient positive charges inside and negative charges outside of the neuronal fibers carry information among neurons.

The release of neurotransmitters and their recognition by receptors on the postsynaptic neuron control the electrical action potentials that transmit the information. The intensity of electrical activity is modulated by neurotransmitters produced in the neuronal cell bodies and transported along axons to their terminals, where they are stored until secreted into the synapse. As the result of electrical action potentials traveling down axons, neurotransmitters are released into the synapses. After the neurotransmitter enters the synapse, it modulates the electrical activity of the postsynaptic neuron.

Networks of neurons in specific regions of the brain receive and transmit information to neurons in other parts of the brain. Hundreds or even thousands of dendrites bring information to every neuron. The frequency of these action potentials—about 1,000 per second—controls the transmission of information throughout the networks of neurons. Molecular messengers (neurotransmitters) bind to receptors to activate or inhibit postsynaptic action potentials.

Neurotransmitters are constantly being synthesized and stored in vesicles at the end of axons. The release of neurotransmitters into the synapse from vesicles in the presynaptic neurons results from the opening of calcium channels. Calcium fuses with the presynaptic neuronal vesicles, which brings about the release of the neurotransmitters into the synapse where they bind to receptors on postsynaptic neurons. Neurotransmitters control emotions and reasoning, affecting thoughts, feelings, dreams, memories, and actions.

Many pharmaceuticals act by inhibiting or enhancing the effects of neurotransmitters, or the action of hormones. Neurotransmitters include DA, 5-HT, NE, and ACh, and their receptors. Hormones include testosterone and estrogens. Drugs affect the synthesis, storage, or release of one or more neurotransmitters or hormones. Antagonist or agonist drugs block or stimulate receptors, respectively.

Some neurotransmitters are small molecules, including ACh, monoamines (NE, DA, and 5-HT), amino acids (primarily glutamic acid, GABA, aspartic acid, and glycine), purines, (adenosine, ATP, GTP, and their derivatives), fatty acids (endogenous cannabinoids), and peptides (vasopressin, somatostatin, and neurotensin). Hormones, such as insulin, have local effects, and circulate in the blood to produce distant effects. Ions, such as zinc, released into synapses, also affect neurotransmission. Neurotransmitters include catecholamines (NE), biogenic amines (5-HT and ACh), polypeptides/proteins (endorphins), and amino acids/amino acid analogs (glutamate and GABA).

D_2-type DA receptors are both pre- and postsynaptic, whereas the D_1-type is postsynaptic. Dean Wong and colleagues at Johns Hopkins found a decrease in DA and 5-HT receptors with age. D_2 DA receptors decreased 6% from 20 to 80 years of age.

A research group at Karolinska Institute in Stockholm has found that women and men differ in the number of binding sites for 5-HT in certain parts of the brain. A doctoral thesis by Hristina Jovanovic showed that women have a greater number of 5-HT receptors than men. Women also have lower levels of the serotonin transporters (SSRIs) on presynaptic neurons. "Men and women sometimes respond differently to treatment with antidepressant drugs," says Anna-Lena Nordström, who led the study.

Some neurotransmitters are second messengers, including cAMP, phosphatidyl inositol (PI_3), protein kinase C, and diacylglycerol (DAG). DA receptors are involved in an elevation of cAMP levels when there is stimulation of D_1 DA receptors, but there is a decrease in cAMP after stimulation of D_2 receptors.

Unbound synaptic neurotransmitters, such as DA, 5-HT, NE, and others are taken back into the neuron by binding to protein transporters on the presynaptic neuronal membranes, which remove neurotransmitters from the synaptic cleft. Neurotransmitters also are inactivated by enzymatic degradation.

"Neurons have to shout together to get a message across and make a reliable signal above the background noise." Three thousand inhibitory and 300 excitatory action potentials per second may be involved in the firing of a single postsynaptic neuron (Churchland, 1988). There are four different types of adrenergic and cholinergic receptors in the human brain, five types of opiate receptors, and 15 types of 5-HT receptors.

DA receptors are abundant in the cortex, hippocampus, striatum, and thalamus. Different subtypes may be expressed by the same neurons. D_2 type are both pre- and postsynaptic, whereas D_1 type are postsynaptic. Researchers at BNL also found a 6% decrease in D_2 DA receptors with each decade of age from 20 to 80 years of age.

Second messengers include cAMP, IP_3, protein kinase C, cAMP and DAG. DA receptors are involved in an elevation of cAMP levels when there is stimulation of D_1 DA receptors, but there is a decrease in cAMP after stimulation of D_2 receptors.

Unbound synaptic neurotransmitters, such as DA, 5-HT, NE, and other neurotransmitters, are taken back into the neuron by binding to protein transporters on the presynaptic neuronal membranes, which remove neurotransmitters from the synaptic cleft. Neurotransmitters are also inactivated by enzymatic degradation.

Neurotransmitters and receptors modify the frequency of action potentials arriving via presynaptic axons depolarize the postsynaptic neurons, activating N-methyl-D-aspartate (NMDA) receptors. In addition to glutamate, the NMDA receptor requires the binding of another molecule, glycine, for the receptor to function. When the neurons are inactive, NMDA receptors are nonresponsive as a result of the blocking of ion channels by magnesium ions.

Calcium influx into the postsynaptic cells induces so-called long-term potentiation (LTP) of postsynaptic neurons. LTP was discovered in the hippocampus in 1973, and it is a mechanism for learning and memory.

6.2. DA Receptors

The imaging of a neuroreceptor in the human brain (DA receptors) by PET in 1983 was stimulated by the work of Creese, Burt, and Snyder (*Science* 192:481–483, 1976), who showed that the antipsychotic effect of drugs in schizophrenic patients was directly related to the degree of blocking of D_2 DA receptors. In 1978, Kuhar and associates at Johns Hopkins showed that DA receptors in the basal ganglia could be visualized by autoradiography after administration of drugs that were bound by DA receptors.

The first successful imaging of a neuroreceptor in the living human brain—the DA receptor—was carried out on May 23, 1983, at Johns Hopkins. Earlier that month, at the Grand Hotel in Saltsjobaden, Sweden, we presented the first results in imaging DA receptors in baboons with PET. The DA receptors were imaged in the caudate nucleus and putamen. with the radioactive tracer C-11 NMSP.

During the presentation of our results in Sweden in early 1983, Marcus Raichle, a neurologist from Washington University in St. Louis sat in the first row as I showed the PET images showing C-11 NMSP bound to DA receptors in the basal ganglia of a baboon's brain. In the discussion period, he said that one had to use fluorine-18 to carry out PET imaging of neuroreceptors in the brain. During the discussion, I responded: "Mark, did you not just see the successful imaging of the receptors with C-11 NMSP?" I cannot recall his reply, but I decided then that I would try to carry out the first PET imaging of neuroreceptors as soon as I returned to Johns Hopkins the following week.

In concluding my presentation, I said, "If we liken the study of the chemistry of the brain to the climbing of a mountain, we have now assessed its height, assembled teams and resources at base camps, and are beginning to see beautiful views. Let us hope we can continue to have the enthusiasm, strength, and resources to continue the ascent." A long climb lay ahead, but cyclotron-produced radiotracers were what made it possible to begin this exciting adventure.

On September 23, 1983, we reported in the journal *Science* (221:1264–1266), "An awake, 56-year old Caucasian male (HNW) was positioned in the PET scanner by the use of a head holder. The middle slice of the PET scan contained a section of the brain including the basal ganglia. The scan was carried out in the same way as had previously been used in the baboons. The images that were obtained 40–60 min after injection and 70–130 min after injection show a high accumulation of C-11 *N*-methyl spiperone tracer activity in the basal ganglia relative to the rest of the brain." We performed a second scan later the same day, after I had been injected with a large dose of nonradioactive spiperone. The subsequent PET images obtained after the "blocking" dose of NMSP showed that the basal ganglia no longer accumulated the tracer, which provided strong evidence that the tracer (C-11 NMSP) was indeed binding to DA receptors.

Neurons containing DA are highly concentrated in the substantia nigra. DA modulates the neuronal activity involved in movement and emotions. These neurons degenerate in patients with Parkinson's disease, decreasing DA transport to the basal ganglia, especially the putamen. The putamen is involved in motor activity; the caudate nucleus in emotions. A major discovery was that the administration of L-dopa, precursor of DA, relieves the symptoms of Parkinson's disease, an example of how molecular imaging can guide molecular therapy.

Since 1983, many neurotransmitters, receptors, and reuptake sites have been imaged to study brain chemistry in health and disease. In 1998, David Townsend and colleagues at the University of Pittsburgh combined CT and positron tomography imaging of the same patient lying on an imaging bed to produce fused images of brain structure and biochemistry.

6.3. Acetylcholine

In 1921, an Austrian scientist, Otto Loewi, discovered ACh, the first neurotransmitter. At the time, it was difficult for neuroscientists to believe that molecules could affect neuronal activity. The principal focus was on electrical action potentials.

ACh, NE, and 5-HT play important roles in the Peri nervous system (PNS), as well as the brain (CNS), with ACh being the neurotransmitter in all the neurons of the autonomic nervous system. Profound biological effects can result from even slight changes of molecular configuration of neurotransmitters.

In the 1950s, Ernst and Elizabeth Florey at the Montreal Neurological Institute discovered the first neurotransmitter, GABA, that had an inhibitory effect on neuronal activity. In contrast, glutamic acid increases neuronal activity.

Joel Elkes and colleagues studied the effects of cholinergic and other drugs, including amphetamines and LSD, on the electrical activity of the brain, and its relationship to behavior. "Rather than thinking in unitary terms, it may, at this stage, be advisable to think in terms of the possible selection by chemical evolution of small families of closely related compounds, which by mutual interplay would govern the phenomena of excitation and inhibition in the central nervous system." Biochemical processes activate and inhibit neuronal networks in regions of the brain involved in visual, verbal, motor, and emotional functions.

The cell bodies of neurons that involve NE are located in the brain stem, with projections to the thalamus, cortex, hippocampus, and cerebellum. Other neurotransmitters, such as ACh and some neuropeptides, are broken down while still in the synaptic cleft. In the synapses connecting neurons and muscle cells, the enzyme acetyl cholinesterase performs this task. Excitatory ACh receptors that activate skeletal muscles at the neuromuscular juncture are called nicotinic receptors because they are stimulated by nicotine.

The rate of change in receptor occupancy is more important than their number. Changes in receptor occupancy can occur in <30 seconds after inhaling cigarette smoke. More and more medications are administered by inhalation, because they bring about a very rapid onset of action.

David E. Kuhl of the University of Michigan, who developed the first clinical tomography machine in 1963, has focused much of his research on the cholinergic system in Alzheimer's disease. In patients with early Alzheimer's, the postero-lateral part of the brain has decreased glucose metabolism, in regions of abnormal ACh synthesis.

The enzyme acetylcholinesterase (AChE) activity in the brain can be measured with PET, by using the ^{11}C-labeled ACh analogs N-[^{11}C]methylpiperidin-4-yl acetate and propionate.

Rinne and colleagues in Finland found decreased acetylcholinesterase activity in the hippocampus in patients with mild cognitive impairment and early Alzheimer's disease. Yet, they concluded that the value of in vivo acetylcholinesterase measurements in detecting the early Alzheimer process is limited. It is not known whether destruction of the cells that make ACh is a cause or a consequence of Alzheimer's disease.

6.4. Hormones

Hormones are circulating molecular messengers, carrying information to cells that affects their activity. Some enter cells, where they bind to intracellular receptors. Estrogen-binding receptors are an example. Among the first image of hormonal receptors was by Elwood Jensen of the University of Chicago in the 1960s, who found that radioactive estrogens accumulated in the uterus of rats. Subsequently, receptors were discovered for other hormones, vitamin D, retinoids, and thyroid hormones.

After hormones bind to receptors in the cell nucleus, they activate appropriate genes, which bring about selective transcription and expression of the hormones. Steroid hormones regulate growth and development, as well as homeostatic processes. Trophic hormones from the pituitary regulate hormonal secretions by the thyroid, adrenals, and gonads.

Hormones are carried throughout the body in large amounts relative to neurotransmitters, often acting slowly over prolonged periods, compared with the fast transmission of information by neurotransmitters, which can occur within milliseconds, a thousandth of the time needed for the action of hormones. For a long time, scientists did not accept the idea of chemical neurotransmission, because they could not believe that chemicals acted fast enough to control neuronal activity. Their attention was focused on the electrical activity in the neuronal networks that was thought to be adequate to account for all of the manifestations of the human mind.

6.5. Molecules of Aggression

PET and SPECT have revealed abnormalities in dementia, movement disorders, depression, anxiety, and substance abuse. Molecular processes also are related to aggressiveness, emotional stability, conscientiousness, sociability, and selfishness.

Thoughts stimulated by controlled external events have been correlated with specific changes in molecular processes, such as F-18 FDG accumulation, in specific regions of the brain. Emotions, such as aggression, also can be correlated to chemical processes in different parts of the brain. Studies of aggressive behavior have focused on the DA, 5-HT, and NE systems, i.e., the monoamine systems.

An example is that from the BNL, led by Nellie Alia Klein and colleagues. PET studies were performed after the intravenous injection of the radioactive tracer C-11 clorgyline, which is bound by the monoamine oxidase (MAO) A enzyme (Figs. 6.6 and 6.7) in different parts of the brain.

The regional concentrations of MAOA in their brains were lowest in those men who were most aggressive. More aggressive men had lower C-11 clorgyline uptake; less aggressive

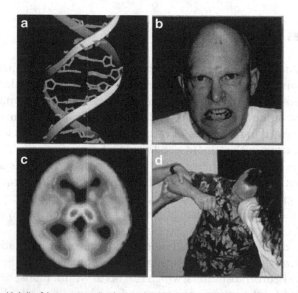

Fig. 6. 6 (a) Section of the double helix of the gene expressing the enzyme MAOA. **(b)** C-11 corgyline PET scan showing the regional distribution in the brain of the MAOA enzyme. **(c)** Aggressive subject. **(d)** A ggressiveb ehavior.

Fig. 6.7 Binding of C-11 corgyline in the brain decreases with increasing aggressiveness.

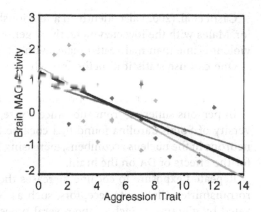

men had higher uptake. Of the 240 questions in the questionaire, only those questions related to short temper, vindictiveness, and enjoying violent movies were related to MAO A levels. There was no correlation between C-11 clorgyline regional uptake in the brain in those persons who were depressed or had other negative feelings. In essence, the MAOA activity in the brain of 27 healthy men was inversely promotional to how aggressive they were.

"The less MAOA the men had in their brain, the more they answered "yes" to statements about taking advantage of others, causing them discomfort ... Our findings corroborate the relevance of brain MAOA in aggressive personality ... If this model of understanding is tested on individuals who engage in violent behavior (such as domestic violence), it may show promise in the future for pharmacological intervention against violence" (Alia-Klein).

The radiotracer C-11 clorgyline binds to and is metabolized by the enzyme MAOA. Clorgyline blocks NE receptors in the brain that, in large doses, results in increased levels of NE in the cerebral cortex. There was no correlation between C-11 clorgyline uptake in the brain in persons who were depressed or had other emotions.

The enzyme MAOA is located in synapses of NE, 5-HT, and DA neurons, which are located in the locus ceruleus, placenta, intestine, and autonomic nervous system. MAOA breaks down 5-HT, NE and DA, rendering them inactive. Inhibition of MAOA activity helps alleviate depression because it slows the breakdown of epinephrine and NE.

The gene that expresses the MAOA enzyme is located on the X chromosome, and it is involved in the control of mood, aggression, and pleasure. Subjects with low MAOA activity react much more strongly to stress than those with high activity.

Ronald Bailey and colleagues at the University of Wisconsin (August 7, 2002) studied a cohort of 442 men from New Zealand whose lives had been followed from birth to age 26. Genotype analysis found 279 subjects had high MAOA activity and 163 had low MAOA activity. Subjects who suffered abuse as children and had low activity of the MAOA gene were nine times more likely to engage in antisocial behavior, such as persistent fighting, bullying, lying, stealing, or disobeying rules in adolescence. "As adults, 85% of the severely maltreated children who also had the gene for low MAOA activity developed antisocial outcomes, such as violent criminal behavior."

Caspi et al. (2002) also identified a relationship between MAOA and anti-social behavior. Males with the low enzyme-activity genotype were more likely to be convicted of a violent crime than maltreated males with a high-activity genotype.

One can use statistical methods, such as Bayes' theorem, to relate neurotransmitters to different types of behavior. It may someday be possible to predict future behavior by measuring regional molecular processes in the brain.

In persons suffering from substance abuse, Cornish and Kalivas of the Medical University of South Carolina found that cocaine blocks the reuptake of DA by presynaptic neurons in the nucleus accumbens, increasing DA levels in the synapse. The addict craves for the effects of DA on the brain.

Repeated exposure to cocaine increases the release of glutamate, the excitatory neurotransmitter. Glutamate receptors, such as NMDA, in postsynaptic neurons, are activated by glutamate. Each of the several types of glutamate receptors is transcribed by different genes and their mRNA.

In addition to being an excitatory neurotransmitter, glutamate is also a precursor in the synthesis of GABA, the most active inhibitory neurotransmitter in the human brain, illustrating the continual balancing of excitatory and inhibitory neuronal activity.

In summary, 5-HT, DA, endorphins, and estrogens are involved in the feelings of pleasure. Epinephrine, NE, and testosterone are involved in pain, depression, and anxiety. High serum testosterone levels are found in aggressive, decisive, tough-minded and competitive people. Men have good spatial perception skills, whereas women with high estrogen levels have good verbal skills.

5-HT-secreting neurons are located in high concentrations in the amygdala and hippocampus (Schroeder et al., 1991). 5-HT is released into the synapse from serotonergic presynaptic neurons, and it is bound to 5-HT receptors on postsynaptic neurons. The unbound 5-HT in the synaptic cleft is taken back up into the presynaptic neurons by so-called protein transporters on these neurons.

5-HT and other monoamines affect emotions, whereas ACh and DA affect movement and emotions. Blocking serotonin transporters on presynaptic neurons with drugs increases synaptic concentrations of 5-HT, which makes depressed patients feel better. Other neurotransmitters, such as endogenous opiates, increase pleasure and decrease pain.

^{11}C raclopride and [^{123}I]iodobenzamine are used to study D_2 DA receptors; ^{11}C cocaine studies DA transporters. Transporters are often targets for treatment of psychiatric disorders, including depression and substance abuse.

6.6. The Sleeping Brain

In 1953, Eugene Aserinsky and Nathaniel Kleitman at the University of Chicago are the founders of modern research on sleep. Using the EEG, Aserkinsky discovered rapid eye movements (REM) that occur is associated when a person is dreaming and is associated with an increase in neuronal activity. The stages of sleep are as follows: stage 1, light sleep; stage 2, an intermediate stage between REM and deep sleep; and stages 3 and 4, deep sleep, when electrical activity is very slow.

During sleep, FDG/PET studies show that there is greatly diminished glucose use throughout the entire human brain. In slow-wave sleep, there is decreased FDG accumulation due to neuronal deactivation of the dorsal pons and mesencephalon, cerebellum, thalami, basal ganglia, basal forebrain, hypothalamus, prefrontal cortex, anterior cingulate cortex and precuneus, and the mesial aspect of the temporal lobe.

Beginning 90 minutes after the onset of sleep, there are periods of intense mental activity during which the closed eyes jerk around under their lids. Persons who were awakened during REM sleep report that they had been dreaming. The first REM period lasts about 10 minutes, and, with each subsequent 90-minute period, the REM period eventually takes up 20–25% of total sleep time. Dreaming is not limited to REM periods. People often recall dreams shortly after falling asleep, during a period of non-REM sleep.

Avi Karni and Dov Sagi at the Weizmann Institute in Israel found that depriving people of REM sleep resulted in impaired memory. Bruce McNaughton and Matthew Wilson at the University of Arizona have linked deep sleep and memory. "When you're asleep, the brain is processing information accumulated when you were awake. It's no longer storing new input; it's organizing information," explains Dr. McNaughton. "Your brain is like a cluttered desktop at the end of the day. At night, when you're asleep and no more information can be put on the desk—or in your brain—your brain can then file away the information."

Sleep is essential for memory formation. When a person is sleep-deprived, the brain's ability to move information from temporary memory to long-term storage is impaired. Information is lost or forgotten.

Sleep improves a person's ability to learn repetitive skills, such as riding a bike or typing. Improvements in learning are connected to REM sleep. When the subjects are deprived on REM sleep, their learning ability is impaired. Canadian researchers compared the performance of students cramming for an examination without sleep with classmates who slept after studying. The students who slept retained more information. "Sleep is a time when the brain can rehearse recently learned material," says James D. Walsh, Ph.D., Director of St. Luke's Hospital Sleep Medicine and Research Center. "If you're sleep-deprived, you'll remember less of newly presented information."

At the Sleep Neuroimaging Research Program at the University of Pittsburgh Medical Center, Eric Nofzinger carried out ^{18}F-FDG PET scans in sleeping volunteers. He found that dreams originate in the limbic system. "That's why so many dreams are emotional events," says Nofzinger, "where we're running from danger or facing an anxious situation. The part of the brain that controls dreams orchestrates our instincts, drives, sexual behavior, and fight or flight responses." The frontal lobes are disengaged during sleep, which may be why dreams are often bizarre combinations of events and people. Patients who suffer from depression have increased REM activity during sleep that originates in the limbic region. ^{18}F-FDG accumulation in the limbic system during REM sleep increases to a greater degree in depressed patients than in normal persons.

6.7. Sleep Deprivation

In studies of normal persons deprived of sleep, and colleagues at the Walter Reed Army Institute of Research, the Maryland Psychiatric Research Center, and the Division of Nuclear Medicine of Johns Hopkins, examined the effect of 24, 48, and 72 h of sleep deprivation on the regional cerebral metabolic rate for glucose (CMRglu), young, normal, healthy male volunteers who were performing complex cognitive tasks. They carried out PET studies with ^{18}F-FDG.

Brain glucose utilization decreases when alertness and cognitive performance are impaired by sleep deprivation, especially in the prefrontal cortex, a region involved in alertness, attention, and higher order cognitive processes and in the thalamus, a subcortical structure involved in alertness and attention.

Absolute and relative regional glucose use decreased after 48 and 72 h of sleep deprivation in the same areas that showed decreases after 24 h of sleep deprivation. The decrease in glucose use in the prefrontal-thalamic network by prolonged sleep deprivation was proportional to the decline in alertness and cognitive performance, whereas the increases in visual and motor areas may reflect the subjects attempt to remain awake and perform assigned tasks. As the subjects tried to fight sleeplessness, characterized by decreased neuronal activity in the prefrontal cortex and thalamus, other regions of the brain increased neuronal activity.

6.8. Chemistry of Sleep

Sleep is associated with increasing concentrations of adenosine in certain parts of the brain. Adenosine is the core of adenosine triphosphate, or ATP, the molecule where energy is stored inside cells. Adenosine is secreted by neurons and glia when neurons are active. The body sleeps to replenish low stores.

Sleepiness is mediated by adenosine receptors. Caffeine, a well-known stimulant in coffee, tea, and some sodas, prevents adenosine from binding to adenosine receptors. It stimulates wakefulness by its antagonistic effect on A1 and A2a adenosine receptors.

The adenosine A1 receptor antagonist [^{18}F]CPFPX is used to carry out PET to study the role of adenosine and its receptors. The urge to sleep is associated with changes in local brain adenosine levels.

Areas of the brain that are active generate adenosine; areas that are not active tend not to generate adenosine (Robert W. Greene, Harvard University). Adenosine concentrations in some parts of the brain increase during waking and decline during sleep. Brain levels of adenosine decline during sleep (Simon M. Ametamey and Peter Achermann, University of Zurich). Adenosine concentrations in the brain rise if animals are forced to stay awake.

In military operations, such as those in Iraq, stimulant drugs that enhance wakefulness are taken to heighten alertness, aggressiveness, and resistance to fear, pain and fatigue.

The Allen Institute for Brain Science in Seattle, WA, sponsors research to study gene expression under five different conditions of sleep and wakefulness. An online database is available describing the 224 genes in the mouse that are associated with sleep. Generated in collaboration with leading sleep researchers at Stanford Research Institute (SRI), this data set is accelerating progress toward understanding and effective treatment of sleep disorders.

6.9. A Word of Caution

"People naturally want to use brain science to inform policy and practice, but our limited knowledge of the brain places extreme limits on the effort. There can be no brain-based education or parenting at this point in the history of neuroscience" (Daniel Siegal, *The*

New York Times, September 17, 2007). Yet, we are making progress. It is a lot easier to study how drugs affect the brain than to discover how the brain works.

Humans are prone to abuse drugs. In persons between the ages of 35 and 54 in the United States, (1) in 2004, there were 18,249 deaths from overdose of illicit drugs; (2) the same from 16,785 fatal inductrial and suicides (3) in 2005, there were > 1 million arrests, 1 million for violent crimes, 500,000 for drugs, and 650,000 for drinking-related offences; (4) there were 21 million binge drinkers (five or more drinks per occasion); and (5) thousands of persons treated in hospital emergency rooms for abusing illegal drugs in 2005.

Methamphetamine (Ecstasy) continues to flow across the Canadian border into the United States in increasing amounts. Seizures of Ecstasy increased 10-fold from 2003 to 2006. Today, the supply of methamphetamine is decreasing (Jane Gross, *The New York Times*, January 9, 2008).

In 2005 and 2006, first-time users of illicit drugs in the United States increased by 40%, one third under the age of 18. Shutting domestic laboratories has substantially reduced the supply of methamphetamine. Yet, first-time users in the United States increased to 860,000 in 2005 from 615,000 in 2006. At the United States' northern border, federal law enforcement officials seized 5,485,619 doses of Ecstasy in 2006, according to the Office of National Drug Control Policy, compared with 568,220 in 2003.

Teenage use of illicit substances has fallen by 23% since the 1990s, and by 50% for LSD and Ecstasy. Of possible relationship is the fall in the birth rate by 35% for 15–19 year olds has since 1991. The high school dropout rate is at a 30-year low.

Marijuana is being harvest in huge amounts in the Canadian province of British Columbia. Recently, the mandatory sentence for growers and traffickers has been increased, and laws are being enforced more vigorously.

More Canadians use marijuana than in any country in Europe, Asia, or Latin America (Douglas Belkin, *Wall Street Journal*, January 8, 2008). More than half of all Canadians think that marijuana should be legalized. Vancouver remains a major source of illicit drugs.

Mayor Sam Sullivan has said, "In my experience, people who consume marijuana are not a problem for public order or crime."

6.10. Addiction Research

In 1983, I wrote, "One of the most intriguing problems in biomedical research today is that of relating manifestations of neuropsychiatric disease to chemical processes in different parts of the brain … Much of what we know about the brain is based on the study of animals. The proper study of mankind is man. Animal models may not reflect normal or diseased human beings. With molecular imaging, we can address the questions: what is happening in the chemistry of the human brain in relation to mental activity? Why is it happening? And how is it happening?

In 1950, when I was a medical student, the tools we had to examine a patient's nervous system included a rubber hammer for testing reflexes, a pin for testing the sensory system, and a wisp of cotton for testing light touch. We would ask questions to evaluate the patient's psychological state. In patients suffering from epileptic seizures, the electroencephalogram could measure the electrical activity of the brain. The treatments available for patients with neurological diseases at that time were vitamin B_{12} for pernicious anemia, vitamins for alcoholics, and antibiotics for meningitis and brain abscess. Psychopharmacology lay in the future.

"The search is being made everywhere to find a chemical marker to detect vulnerability to disease before deterioration of the brain has begun" (Moorhead, 1983). Today, PET studies of brain glucose utilization and the chemistry of neurotransmission are playing an ever-increasing role in developing more selective therapeutic agents for treating a variety of mental illnesses.

To study the effects of drugs, Sydney Udenfriend and Bernard Brodie developed a method for the determination of 5-HT in body tissues. The procedure uses a spectro-photo-fluorometer to activate and measure emitted fluorescence with a frequency from 250 to 650 μm. 5-HT is activated maximally at 295 μm and emits fluorescent light with a maximum at 550 μm.

In 1890 the philosopher, William James, said that, "Chemical action must of course accompany brain activity but little is known of its exact nature." The spectrophotofluorometer made it possible to begin to link brain chemistry and behavior.

Among its first applications was the study of reserpine. Reserpine was isolated in 1952 from *Rauwolfia serpentina* (Indian snakeroot), which was used for centuries to treat insanity. Reserpine acts by blocking the reuptake NE back into presynaptic vesicles. The same year, Henri Laborit, a surgeon in Paris, administered reserpine to prepare his patients for surgery, because it maden them less anxious. Subsequently, it was widely used to treat patients with severe psychiatry diseases with impressive results.

Chlorpromazine was the prototype of antipsychotic drugs of the phenothiazine class of drugs. In March, 1954, the drug, under the name Thorazine, was approved by the Food and Drug Administration (FDA). It blocks a variety of receptors, including the cholinergic, dopaminergic, and histamine, as well as adrenergic receptors. Along with many of the older antipsychotic drugs, it has little effect on the serotonergic system.

Chlorpromazine reduces DA activity by blocking mesolimbic DA receptors. Postmortem studies have shown high DA concentrations in subcortical brain regions and greater than normal DA receptor densities in the brains of schizophrenic patients.

The negative/deficit symptoms of schizophrenia are associated with low DA activity in the prefrontal cortex. Positive symptoms, such as hallucinations and delusions in schizophrenia, are related to excessive DA activity in mesolimbic DA neurons.

The drug company, Smith Kline & French, first marketed chlorpromazine as a treatment for vomiting. Clinical trials in patients with mental illness were then conducted in state mental hospitals by Pierre Deniker who obtained "miraculous" results. In 1954, two years after chlorpromazine, reserpine was introduced into clinical practice.

The three monoamine neurotransmitters—DA, 5-HT, and epinephrine—are helpful in patients with depression. Amphetamines relieve depression by stimulating epinephrine release. Blocking the reuptake of DA by presynaptic neurons with DA-reuptake inhibitors, such as bupropion and amineptine, is the mechanism of relieving depression. Blocking postsynaptic D_2 DA receptors results in the relapse of patients who have achieved remission from depression. This blockade can also counteract the effectiveness of SSRI medications.

Measuring the effects of molecular processes on symptoms in patients with mental disorders does not provide a biological explanation of what "causes" these diseases, but rather provides modifiable molecular manifestations of these disorders. The emergent mind is strongly affected by molecular processes in the brain.

We will never find a biological cause of love, hate, anger, or violence, but we can identify molecular processes, such as those involving DA, 5-HT, epinephrine, and other molecules that affect the brain.

It is a lot easier to learn how drugs affect the brain than to learn how the brain works. Molecular imaging lets us examine the molecular processes affected by drugs. It might even be possible to develop a molecular model of violence, war, and peace.

The model should be (1) descriptive, (2) predictive, and (3) prescriptive, leading to actions. Francis Bacon in *Novum Organum* wrote, "Nature to be commanded must be obeyed ... The true method of experience first lights the candle (i.e., generates an hypothesis), and then, by means of the candle, shows the way, and then experiment itself shall judge."

Many of the 80 million baby boomers born in the United States between 1946 and 1964 have had their lives affected by drugs, including Ritalin, Prozac, Lipton, and Elvira. We need better understanding of the biochemistry of these drugs and their effects on behavior. Well-planned, ethical research relating brain chemistry and behavior can be carried out in the criminal justice system, although some claim that prisoners can never give truly informed consent to such research.

Once it was thought that witches made men and women love or hate. Shakespeare wrote, "There are more things in heaven and earth that can be dreamt of in your philosophy ... There is no good or bad than thinking makes it so."

Hamlet said, "There's a divinity that shapes our ends, rough hew them how we will." William Harvey wrote that "Every affection of the mind that is attended with either pain or pleasure, hope or fear, is the cause of an agitation whose influence extends to the heart" (Excitatio anatomica de motu cordis e sanguinis in animalibus. 1628).

Tracers that have been developed to examine postsynaptic D_2 DA receptors include ^{11}C-raclopride, ^{18}F-fluoroethylspiperone, ^{11}C-NMSP, and ^{123}I-epinephrine. The choice of tracer depends on whether one wishes to examine extrastriatal dopaminergic regions, release of endogenous DA, or the total number of D_2 DA receptors in a specific region.

In adrenergic neurons, L-dopa is converted to NE by DA β-hydroxylase or to epinephrine by phenyl-ethanolamine-N-methyl transferase. The rate of synthesis of DA is determined by the amount of available tyrosine hydroxylase, and is a key factor in determining the rate of neuronal firing or inhibition by DA. DA is metabolized by the enzyme catechol-O-methyl transferase and MAO.

After release from storage sites in presynaptic monoamine transporter vesicles, DA is bound by postsynaptic DA receptors, and unbound DA is broken down by enzymes, or taken back into presynaptic neurons, called "reuptake" into presynaptic neurons by specific presynaptic protein transporters. The uptake of neurotransmitters by presynaptic neurons is a major mechanism for stopping neurotransmission. The recycled neurotransmitter is repackaged in vesicles in presynaptic neurons, and then released again in response to stimulation by presynaptic electrical action potentials. Tracers have been developed to assess the vesicular membranes, as well as the specific reuptake protein transporters. The selectivity with which monoaminergic neurons store DA, 5-HT, NER, or histamine in presynaptic vesicles depends on the specificity of the membrane transporters.

Abnormalities in neurotransmission are found in many serious neurological diseases, including dementia and Parkinson's disease. Fluorine-18 dopa was first used as a tracer for dopaminergic presynaptic neurons by Steve Garnett in 1983. Measurement of ^{18}F-dopa in the striatum can be in the early detection of Parkinson's disease. There is diminished uptake of ^{18}F- dopa in the striatum of these patients. A large number of psychoactive drugs affect monoamine neurotransmitter systems: 5-HT, DA, and epinephrine (Glennon et al., 1984; Ismaiel et al., 1993; Winter et al., 1999). Pharmaceutical companies are continually trying to develop better drugs to affect thinking, feeling, and behavior, drugs that are safer and more effective. For example, in schizophrenic patients, or in persons abusing amphetamine or cocaine, phenothiazines block DA receptors and reduce psychotic symptoms.

The tracer, ^{11}C-raclopride binds to D_2-DA receptors. Decreased accumulation of the tracer results from the release of endogenous DA that competes for binding with the tracer. Jon-Kar Zubieta and colleagues at the University of Michigan showed that the extent to which a person responds to placebos correlates with DA release in the nucleus accumbens. These persons who responded to placebos had more DA-related neuronal activity in their left nucleus accumbens than other volunteers (*Neuron*, July 19, 2007). DA release in the nucleus accumbens is involved not only in the ability to experience pleasure and reward, but also related to the "high" caused by illicit drugs. DA release is increased when the person anticipates pleasure.

Early studies indicated that DA was increased in patients with schizophrenia. There was increased availability of DA receptors in patients with schizophrenia, but later experiments in Sweden were not able to confirm this finding. The DA hypothesis has been challenged. People with schizophrenia are not the only patients who respond to antipsychotic medication. Antipsychotic medication does not significantly affect the negative symptoms of schizophrenia. There is more to schizophrenia that just abnormal DA effects.

However, increased amounts of synaptic DA in the striatum are responsible for "positive" symptoms, such as paranoia and increased motor activity. The prefrontal cortex also has been implicated in causing the lack of perception, hallucinations, and delusions. D_1 DA receptors in the prefrontal cortex are involved in the negative symptoms of schizophrenia.

The limbic system has particularly high concentrations of D_3 and D_4 receptors, which are involved in the production and inhibition of emotions. The abnormally low dopaminergic activity in the prefrontal cortex and limbic system result in a decrease in inhibition in the production of dopaminergic neuronal pathways.

When DA binds to postsynaptic receptors, there is an opening of axonal sodium and potassium ion channels that bring about the movement of electrical action potentials along postsynaptic axons. In 1972, Paul Greengard of Yale University showed that DA also increases a second messenger, cAMP. The relative potencies of phenothiazines in blocking DA's effects on cAMP are correlated with their ability to alleviate symptoms of patients with schizophrenia. Butyrophenone drugs are more potent than phenothiazines. Haloperidol is 10–20 times more potent, whereas spiroperidol is 10–20 times more potent than haloperidol. All of these drugs block the binding of DA by D_2 DA receptors (*Brainstorming*, Solomon Snyder, Harvard University Press, 1989).

Amphetamines and cocaine stimulate the release of DAe from presynaptic storage vesicles. DA acts on the nucleus accumbens and striatum when a person is stimulated by rewarding experiences, such as eating or sex. Cocaine blocks the reuptake of DA by DA receptors on postsynaptic neurons, which leads to increased levels (up to 150%) of

DA in the synapse. Drugs that reduce DA activity (antipsychotics) reduce motivation and inhibit the ability to experience pleasure.

DA is also produced by the neurons in the hypothalamic nucleus, and is secreted into the hypothalamic/hypophyseal circulation. It inhibits the secretion of prolactin from the anterior pituitary gland. DA acts as a modulator of neuronal activity. If dropped directly on isolated neurons, it has an inhibitory action on action potentials.

6.11. Violence and Aggression

Aggression has been related to specific biochemical processes in the brain. Can this type of research now be expanded to include the study of violence, assault, rape, robbery, homicide, suicide, child and elder abuse, and war? Should we search for "modifiable molecular manifestations" of these undesirable human traits? Should not biology join psychological, cultural, and political studies of war, to try to find out whether understanding of molecular processes in the brain might someday help decrease violence and war?

Beatings, gunshot wounds, and stabbings are clearly related to drugs and alcohol abuse. Drug traffickers and street dealers will go to any length, no matter how violent their behavior, to make money. Social factors include the prevalence of hostile environments, where children raise themselves on the streets, and view society as miserable and violent. They must be taught to accept responsibilities, and be respectful of authorities in order to escape the world of constant fear, terror, ignorance and envy. Without moral principles, they can never be free.

A better understanding of brain chemistry and its relation to behavior may facilitate addressing these societal factors. We can try to find out the influence of brain chemistry in producing violent behavior. We can detect regional biochemical difference in the brain between patients with depression and dementia (Fig 6.8).

In Parkinson's disease, there is degeneration of neurons in the substantia nigra. DA is deficient in the basal ganglia, especially the putamen, but also in the caudate nucleus. Administration of the DA precursor, L-dopa, to patients with Parkinson's disease increases DA synthesis, decreasing their difficulty in moving and stiffness, a striking example of how therapy can be based on detecting molecular abnormalities. Dopaminergic neurons in the frontal lobes are also involved in Parkinson's disease, resulting in difficulty in memory, attention, and problem solving. This shows how there is no clear-cut boundary between Parkinson's disease, and Alzheimer's disease. Problems are created by putting a patient in a single "diagnostic" category. Recently, efforts are being made to better characterize what is likely to be the patient's response to drug treatment. This is called personalized medicine.

Molecular imaging can examine the quantitative effects of drugs given to affect perception, learning, movement, memory, hallucinations, paranoia, anxiety, fear, rage, greed, arrogance, aggression, reward and punishment. Many drugs involve the dopaminergic, serotonergic, cholinergic, and opiate neurotransmitter systems. For example, scopolamine and ACh affect language and memory. Hormones, including testosterone and estrogens, affect emotions, including aggressiveness and violent behavior.

PET and SPECT studies of the human brain are now widely used by the pharmaceutical industry in drug design, development, and regulatory approval. Drugs that bind to neurotransmitter receptors can be labeled with ^{18}F- or ^{11}C-, and their distribution was

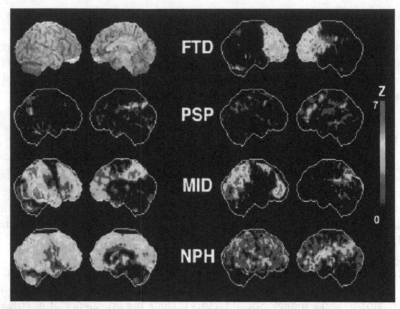

Fig. 6. 8 Statistical images showing the [18]F- FDG images of glucose utilization in the brains of patients with frontotemporal dementia (FTD); progressive supranuclear palsy (PSP); mild dementia of Alzheimer type (MID): and normal pressure hydrocephalus (NPH).

first imaged in mice, rats, or baboons; and then in normal humans; and eventually in patients with mental disorders, such as schizophrenia or depression. Biomarkers used by (1) FDA; (2) NIH; (3) academia; and (4) industry are based on assessing metabolism, cell proliferation, apoptosis, angiogenesis, cellular invasion, and intra- and intercellular communication.

Companies, including Siemens, General Electric (GE), and IBA, provide SPECT/CT and PET/CT instruments and cyclotrons for hospitals, universities, and pharmaceutical research laboratories. For example, at Yale University, there is a cyclotron, a GE 16 slice scanner, two GE 4 slice scanners, and one GE single slice scanner, all with helical capability. A 64-slice scanner was built in February 2006. The 64 slice scanner replaced a 4 slice scanner in the Emergency Department in October 2006.

Today, IBA's mini-cyclotrons are used all over the world. The most common products are ^{18}O to provide ^{18}F; ^{13}N to provide ^{11}C; and ^{82}Rb to provide ^{68}Ga. In the latter case, the use of generators makes it possible to carry out studies away from the cyclotron. Three to 4 Curies of ^{18}F- can be produced routinely by automated synthesis systems. ^{11}C-carbon monoxide and ^{11}C-carbon dioxide are used in the labeling of many molecules.

In the future, small tabletop cyclotrons will be in every hospital or research laboratory together with kit-based microchemistry that allows rapid production of single doses of ^{18}F- and ^{11}C-radiotracers. Single dose production systems are based on microfluidic chemistry, which enables tiny drops of fluids to be manipulated on a silicon chip. A polytetrafluoroethylene matrix provides the basic element. Doses can be synthesized in one chemical step in 5–10 minutes in platforms producing a single tracer (Ron Nutt, Advanced Biomarker Technologies). We live in the Age of Simplification. Unfortunately, today the cost of development of a new radiotracer is close to $20 million each. In-house preparation of tracers will supplement the efforts of the 100 regional laboratories in the United States that provide ^{18}F- FDG to hospitals.

6.12. Images, Words, and Numbers

Artists create images without words, but medical practice needs images, words, and numbers. Effective treatment depends on information transformed by the brain into knowledge. Words reflect the past and the present. David Hume wrote, "To hate, to love, to think, to feel, to see; all this is nothing but to perceive ... All of our simple ideas in their first appearance are derived from simple impressions. We cannot have an idea of the taste of a pineapple without having actually tasted one." In caring for patients, the physician's experience is needed to provide excellent care. Information begins with listening to the patient, then with that obtained from physical examination. A frequent manifestation is pain. Another is a depressed appearance. A patient is depressed when he talks about suicidal thoughts. We don't see or hear the disease, but only the manifestations of the disease.

Physicians gain knowledge about disease by drawing on his perceptions of previous patients and those perceptions of others. What we would like to know is how information, such as the manifestations of disease, is recorded and stored in the physician's brain. How does our brain encode these manifestations? Is there a perception code analogous to the code of nucleotides that are the basis of the genetic code? Is there a perception code of the mind related to the molecules in the brain? What is this molecular code? Are perceptions of external reality encoded in the spatial patterns of neurons? Is it even possible that the perception code of neurons in the brain is based on the same four-letter code contained in genes?

Molecules contain an enormous amount of information. For example, removal of a single functional group from the neuronal stimulant, glutamic acid, transforms the molecule into a neuronal inhibitor, GABA. Another example is the effect of steroidal sex hormones on hypothalamic neuronal networks involved in reproductive behavior and physiology. Estrogens affect cognitive functions, such as memory, by their effect on the hippocampus.

Both short and long-term memories involve phosphorylation of proteins in synapses. In studies of the sea slug *Aplesia*, Eric Kandell found that a mechanism of short-term memory is protein synthesis in specific synapses, leading to synaptic growth. A polyadenylation element-binding protein in the cytoplasm is one of these proteins. Long-term memory results from the creation of specific, selective neural pathways that store information that can be recalled weeks, months or even years later. For example, the temporal lobe is thought to be involved in the long-term memory of visual images.

6.13. Molecular Medicine

Many disorders are the result of biochemical or physiological processes becoming exaggerated in some or deficient in others. For example, pheochromocytomas release large quantities of adrenaline, causing hypertension, anxiety, episodes of sweating, and headaches.

Other manifestations of abnormal molecular processes in the brain include tachycardia, palpitations, tremors, chest pain, nausea, weight loss, and heat intolerance.

Pheochromocytomas can be detected by SPECT imaging of the adrenals with [^{123}I] meta-iodobenzylguanidine (MIBG), which accumulates in cells producing NE or adrenaline. The tracer is taken up into the vesicles of sympathetic nerve endings and subsequently secreted.

Guller and colleagues from Switzerland, collaborating with investigators at Duke University Medical Center and other investigators in North Carolina, studied 152 patients with pheochromocytomas. The most sensitive tests for detecting the disease are total urinary nor-metanephrine (96.9% abnormal), platelet NE (93.8% abnormal), and MIBG scintigraphy (83.7% abnormal). The latter technique could tell where the tumors were located. MIBG scintigraphy, combined with measurement of platelet NE, had a sensitivity of 100%.

In patients with Parkinson's disease, there is diminished accumulation of [^{123}I]MIBG in the apex and inferior wall of the left ventricle, as a result of denervation of sympathetic nerves in these regions.

In 1955, *Bernard Brodie* and colleagues at the NIH discovered that the drug, reserpine, results in a decrease in serotonin in the brain. Reserpine was isolated in 1952 from the dried root of *Rauwolfia Serpentina* (Indian snakeroot), and introduced into clinical medicine in1954. Reserpine blocks the uptake and storage of NE and DA into synaptic vesicles by inhibiting the vesicular monoamine transporters.

Arvid Carlsson was working with *Brodie* at that time in research involving the neurotransmitter, 5-HT. Amines and peptides released from axons at the end of nerve terminals cross synapses and activate or inhibit the electrical activity of postsynaptic neurons by binding to receptor proteins on the postsynaptic neurons. It was found that DA, 5-HT, GABA, benzodiazepine, acetylcholine, opioid, epinephrine, and NE are bound by receptors. Epinephrine is part of a class of molecules known as catecholamines that regulate the activity of the sympathetic nervous system.

Drugs, including ephedrine, stimulate the sympathetic nervous system. Agonist drugs that bind to adrenergic receptors mimic the actions of NE and epinephrine. In addition to binding to receptors, they facilitate the release of epinephrine, NE, or both. β1-Adrenoreceptors are present in cardiac muscle, where they are responsible for the force and rate of ventricular contractions. β2-Adrenoreceptors are found in smooth (bronchial, vascular, gastrointestinal, and genitourinary) and skeletal muscle.

Maintaining blood pressure within a narrow range is one of the major functions of the autonomic nervous system. The symptom, called orthostatic hypotension was described in 1826 by the distinguished French physician, Pierre Piorry. He was called to see a patient who had lost consciousness. His pulse was feeble, and he had to be supported in a sitting position by his friends. When Piorry laid him down, he immediately opened his eyes, his breathing became normal, and color came back to his face. He was suffering from a precipitous fall in blood pressure whenever he stood up.

In 1868, Thomas Addison, described a similar patient: "For the last month he had had attacks of dizziness and dimness of sight, accompanied by a peculiar pain in the back of the head and loss of consciousness. These attacks would always occur when in the standing position and were instantly relieved by sitting or lying down."

When a normal person stands up, blood vessels throughout the body must constrict, increasing vascular resistance that results in a decreased blood flow to all vascular beds except the brain. Molecular responses include the release of antidiuretic hormone (ADH) from the posterior pituitary, and the release of NE and epinephrine from the sympathetic nerves and adrenal medulla.

Even "normal" persons, particularly older persons, who have been in bed for a long time, become faint and dizzy when they first get up. When this phenomenon is severe, it has been called effort syndrome, or neurocirculatory asthenia.

Burst et al. in 1992 examined the effect of vasopressin (ADH) on ^{11}C-methionine accumulation in the brain of living dogs. When nonradioactive phenylalanine was given before the dose of ^{11}C-methionine, the rate of transport of methionine decreased because of competitive inhibition.

The purpose of the study was to test the hypothesis that vasopressin's effect on memory might be through amino acid transport because vasopressin itself does not cross the blood-brain barrier. Vasopressin administration inhibited ^{11}C-methionine transport.

The circulatory system is controlled by epinephrine released into the circulation from the adrenal medulla, and NE released from nerve endings of neurons projecting from the locus ceruleus to the cerebral cortex, limbic system, and spinal cord. NE stimulates the conversion of glycogen to glucose in the liver, increases conversion of fats to fatty acids, and relaxes bronchial smooth muscles. These actions are part of the body's response to stress in what has been called the flight or fight response.

People differ in their resting levels of epinephrine and NE and the amount released in response to stress. Epinephrine provides the increased energy needed during exercise and emotional states, not only when a person is afraid or angry, but also when he or she is excited, anxious, jealous, or happy. Both epinephrine and NE are released when someone is faced with something unfamiliar or intrusive. The heart rate and cardiac output increase, coronary arteries dilate, blood vessels in the intestines constrict, digestion stops, and attention, learning ability, memory, and concentration increase.

Drugs, such as propranolol, block NE receptors, slowing the heart rate. James Black first developed the blockers of adrenergic receptors, which relieved the chest pain of patients with angina pectoris. He received the Nobel Prize in 1988, because these drugs revolutionized the treatment of angina pectoris. Another adrenergic receptor is the α receptor, which brings about vasoconstriction. Antagonist drugs that block α are used to treat hypertension.

During the summers of 1948–1951, I began to carry out research on the autonomic nervous system working in the laboratory of Curt Richter, Professor of Psychobiology at Johns Hopkins. He introduced me to the work of his heroes Claude Bernard, Walter Cannon in Boston and Hans Selye in Montreal. Dr. Richter took me to visit Selye, an endocrinologist in the Biochemistry Department at McGill University.

In 1932, after studying in Prague, Paris, and Rome, Selye had come to Johns Hopkins and became a close friend of Richter. In his research focusing on the relationship between behavior and the endocrine system, Richter had become a skilled surgeon of the adrenal and other endocrine glands of rats, both wild and domesticated. He operated through a microscope long before this instrument had become an important part of human surgery. Julius Jacobson, a classmate of mine in the 1952 Johns Hopkins medical school class, subsequently pioneered the use of a microscope at surgery to repair diseased blood vessels.

In 1945, Selye had become the first Director of the Institute of Experimental Medicine and Surgery at the University of Montreal, where he served until his retirement in 1976.

He observed that stress in rats led to enlargement of the adrenal cortex, atrophy of the thymus, spleen, and lymph nodes, and ulcers of the stomach and duodenum.

Ultrasound is used chiefly to measure blood velocity, but recently research is based on the use of microbubbles to serve as an intravascular contrast agent to image blood flow at the level of the microcirculation, e.g., in the myocardium.

MRI can produce exquisite images of body structure with a near cellular resolution of about 10 μm. One can inject nontoxic, membrane-impermeable tracers to obtain high-resolution images with molecular probes. These are activated in vivo, making it possible to obtain molecular MRI with greater spatial and temporal resolution than is possible with PET. Two types of studies are being studied: enzymatic processing of MRI contrast agents; and binding of the tracer to intracellular receptors. These chaperones bind to specifically targeted enzymes (Thomas Meade, Northwestern University).

Attenuation and scattering in the body are problems in optical imaging. With near infrared (NIR) light, one can obtain 1-mm spatial resolution. In noninvasive imaging of the human brain, the spatial resolution of NIR today is equal to that of autoradiography. The spatial resolution of MRI is approximately 1–200 μm, which provides remarkable structural detail. Brain studies with NIR tracers are possible in the human brain, because sufficient numbers of photons can penetrate the adult skull.

Mental activity involves adenosine, angiotensin, atrial natriuretic factor, bombesin, bradykinin, calcitonin, calcitonin-gene related peptide (CGRP), adrenomedullin, cannabinoid, cholecystokinin and gastrin, corticotropin-releasing factor (CRF), glutamate, histamine, melatonin, somatostatin, thyrotropin-releasing hormone, and vasopressin and oxytocin.

Bombesin is a peptide found in the neurons controlling the gastrointestinal tract by stimulating release of gastrin and pancreatic enzymes and causing contraction of the gallbladder. Cholecystokinin, a hormone secreted by the mucosa of the intestine, has similar effects. Bombesin plays an important role in the control of food intake, and may contribute to anorexia, bulimia, obesity and depression. It has several forms: gastrin-releasing peptide and/or neuromedin B, each with specific receptor subtypes. Bombesin is released from the gastrointestinal tract in response to ingested food, and inhibits appetite.

Among disorders affecting gastrointestinal function are anorexia nervosa and bulimia nervosa, thought to occur in persons lacking self-esteem. Neurotransmitter deficits have been found in patients with bulimia, as a result of vomiting.

K. Herholz and colleagues measured regional cerebral glucose metabolism with PET in patients with anorexia nervosa. During the anorectic state, significant caudate hypermetabolism was found bilaterally (*Biol. Psychiatry.* Jan 1987).

C. M. Gordon and colleagues from Children's Hospital, Boston, MA, carried out PET measurements of regional cerebral blood flow in female patients with anorexia nervosa (*J. Pediatrics* 139:11, July 2001). They were found to have elevated blood flow in both medial temporal lobes compared with control subjects. They also had greater neuronal activation within the left occipital cortex and right temporal-occipital cortex. They postulated that patients with anorexia nervosa have an exaggerated response in the visual association cortex when confronted with large amounts of food. In patients with bipolar disorder, MRI studies has found that the amygdala is smaller in children with bipolar than in healthy children. This has been proposed as having a relationship to the fact that children with bipolar disease have trouble identifying facial expressions.

Dr. Ellen Leibenluft, an NIH researcher, wrote, "Our results suggest that children with bipolar disorder see emotion where other people don't. They also suggest that bipolar disorder likely stems from impaired development of specific brain circuits, as is thought to occur in schizophrenia and other mental illnesses."

Leibenluft and colleagues found differences in the amygdala in a study of 22 children with bipolar disorder. The left amygdala was more activated in children with bipolar disorder than in controls when the subjects were asked to rate the hostility of an emotionally neutral or hostile face (*Proc. Nalt. Acad.. Sci. USA*, June 6, 2006). Compared with controls, bipolar patients perceived greater hostility in neutral faces and reported more fear when viewing them. Such a face-processing deficit could help account for the poor social skills, aggression, and irritability that characterize the disorder in children. Identifying regional abnormalities can help focus attention in studies of the underlying molecular processes related to behavior.

Bradykinin is a potent vasodilator that causes contraction of nonvascular smooth muscle, increases vascular permeability and it is also involved in the mechanism of pain. It has similar actions to histamine, and, like histamine, it is released from venules rather than arterioles. Bradykinin stimulates nitric oxide (NO) production by vascular endothelium. Histamine, released from mast cells in response to injury, inflammation or allergic responses, causes arteriolar vasodilatation, venous constriction in some vascular beds, and increased capillary permeability.

Histamine was found to be a neurotransmitter in the brain in 1982, by T. Watanabe from Tohoku University and H. Wada from Osaka University. It was associated with a presynaptic receptor, with methyl histamine acting as an agonist, and thioperamide as an antagonist, The histaminergic neuronal system has the same distribution in the body as the noradrenergic and serotonergic systems, and it is involved in memory, emotions, and behavior. No disease has yet been found to be due to a disturbance of the histaminergic neuronal system.

Calcitonin, produced in the thyroid gland, is involved in calcium and phosphorus metabolism. Endogenous substances such as glucocorticoids, NO, nerve growth factors, and steroid hormones modulate CGRP release. CGRP receptors are involved in vasodilatation and smooth muscle relaxation, acting to increase cAMP levels and diminish the release of acetylcholinesterase. In vascular smooth muscle, elevated cAMP levels decrease vascular tone. CGRP also activates acetylcholine receptors in the brain.

Japanese scientists found a peptide called adrenomedullin that could raise platelet cAMP levels. Plasma adrenomedullin is elevated in many cardiovascular disorders, and plays a role in controlling blood pressure.

Cannabinoid (CB) receptors are distributed throughout the brain, with high levels in the hippocampus, and they may be involved in storage of newly acquired information or affect changes of mood and behavior. Outside the brain, CB1 receptors are found in the testes, endothelial cells, and smooth muscles of the ileum.

In the 1950s, Guillemin and Rosenberg, and Saffran and Schally discovered that hypothalamic extracts that could stimulate the release of ACTH from anterior pituitary cells. This represented another connection between the brain and the endocrine system. CRF is the major regulator of the baseline and stress-induced release of ACTH, β-endorphin, and other proopiomelanocortin-derived peptides from the anterior pituitary.

The anterior lobe of the pituitary secretes growth hormone, thyroid-stimulating hormone, adrenocorticotropic hormone or ACTH, follicle-stimulating hormone and luteinizing hormones, melanocyte-stimulating hormone, and endorphins.

Two hormones, the anti-diuretic hormone (ADH) and oxytocin, are secreted into the blood by the posterior pituitary gland. ADH raises blood pressure by constricting capillaries, and it reduces urine flow by causing reabsorption of water by the kidney tubules. It also causes contraction of the smooth muscle of the gastrointestinal tract. The effect on the smooth muscle of blood vessels and the intestine is not antagonized by adrenergic blocking agents nor prevented by denervation of the blood vessels. Pitressin is used to treat diabetes insipidus and to prevent postoperative abdominal distention and dispel interfering gas shadows in radiographs of the abdomen.

7

The Physics of Radioactivity

In Paris in 1896, Henri Becquerel, a physicist born in Paris in 1852, was studying fluorescence, the phenomenon where certain minerals, after exposure to sunlight, emit a faint luminosity when in the dark. Both his father and grandfather were physicists. The year before, he had been appointed Professor of Applied Physics. William Conrad Roentgen in Germany had described electromagnetic radiation that has a shorter wavelength than visible light. Four months later, Becquerel thought that this radiation might be related to the glowing light produced by uranium salts that he had inherited from his father, and glowed when exposed to light.

He placed these uranium salts on a photographic plate covered with opaque paper, exposed the covered plates to sunlight, and then developed them. To his delight, he saw that, despite the opaque paper, the photographic plate had darkened at a spot just under the uranium salt. Then, chance favored his prepared mind. He left a packet of uranium salts on a photographic plate in a drawer without exposing the plate to the sun as he had always done in previous experiments. He took the plate out of the drawer, without exposing it to the sun, and developed it. The photographic plate was fogged. He realized that what had darkened the plate must have come from the uranium salt itself. He had discovered natural radioactivity, a term later coined by his graduate student, Marie Curie.

Roentgen had discovered what he called "x-rays," emitted from a cathode ray tube. His discovery and subsequent experiments were the topic of over 1,000 articles and more than 50 books on the subject of x-rays in 1896.

Marie Curie had just arrived from Poland when Becquerel made his revolutionary discovery. Professor Becquerel did not see the importance of his momentous observation, and asked young Marie to try to find out what was going on. Working with her new husband, she did not accept Becquerel's conviction that uranium was a unique fluorescent substance. She systematically examined all kinds of minerals to see if they gave off the penetrating rays. She obtained tons of pitchblende from a mine in Joachimsthal, Czechoslovakia. Working in a large shed in Paris, she and her husband stirred and boiled cauldrons of ore residues and solvents, reducing kilograms of material into a few grams of distillates. Her tiny extracts continued to glow intensely.

She found that all uranium salts were active, but so was *aeschnite*, which contained no uranium. Pitchblende produced emanations of greater intensity than could be attributed to its uranium content.

In 1902, after 4 years of back-breaking efforts, Marie Curie announced that she had extracted a gram of what she called *radium*, from ten tons of starting material. It seemed to continually pour out energy that was able to expose photographic film without any decrease in weight, or change in any other measurable way. Marie and her physicist husband had no idea that a small part of the mass of the material was being converted to high-energy particles and photons. They did realize that they had added a new element to the periodic table that had been proposed 30 years before by the Russian chemist, Dmitri Mendeleyev. He found that, when all the known chemical elements were arranged in order of increasing atomic weight, the resulting table displayed a recurring pattern, or periodicity.

In 1898, in Cambridge, England, a New Zealander, Ernest Rutherford, demonstrated that there were at least two different types of radiation with different penetrating power. He called these *alpha* and *beta* radiation. He subsequently worked at McGill University in Montreal, Canada, and found more radioactive elements: different types of radium and thorium, and actinium. He proposed that these were links in chains of radioactive materials, called the transformation theory. Rutherford and his colleague, Frederic Soddy, described that the rate of decay of radioactive elements were characteristic of the element, and came to be known as half-life. Decay follows the law of probability. Over a given period of time, each atom has a certain probability of decaying, a process that results from the random movements of the subatomic components of the radioactive atoms. This was the first instance in physics of a truly unpredictable phenomenon. The decay of a radioactive atom was probabilistic.

In 1905, Albert Einstein had shown that matter could be related to energy by the most famous equation of all times: $E = mc^2$ where E represents energy, m represents mass, and c^2 is a huge number, the square of the speed of light. The formula shows that mass and energy can be converted into each other, and are related by the speed of light in a vacuum. Nothing can travel faster than the speed of light. One can write Einstein's equation in the form: $E/m = c^2$, which states that the amount of energy in matter is enormous.

In 1903, Marie Curie, her husband and Henri Becquerel received the Nobel Prize in physics; Marie won another Nobel prize (chemistry) in 1911. In 1900, Max Planck had postulated that light energy must be emitted and absorbed in discrete particles, called *quanta*. In Paris in 1924, Victor de Broglie concluded that if light could act as if it were a stream of particles, particles could have the properties of waves. Both quanta and waves are central to *quantum physics*. Quantum theory states that energy comes in discrete packets, called *quanta,* which travel in waves. The principle of *wave-particle duality* states that all subatomic particles can be considered as either waves or particles. Light is a stream of *photon* particles that travel in waves.

In 1916 Einstein hypothesized that electrons can jump from one orbit surrounding the nucleus to another, the jumps occurring in a model of the atom proposed by Niels Bohr. The jumps follow the law of probability.

In 1928, the Russian physicist, George Gamow, and two American physicists, Edward Condon and Ronald Guerney proposed separately that *alpha* particles, identical to the nuclei of helium atoms, were present inside heavy, unstable nuclei because a strong force kept them within the nucleus. According to the laws of probability, occasionally one of them reaches the edge of the atom, and flies away. When quantum mechanics was developed, its recognition of the existence of variance and uncertainty was very controversial when its theories were presented to physicists and philosophers. When it was finally accepted, it answered many questions, including the conduction of heat and electricity, and transparency to light.

In 1913, Neils Bohr applied quantum theory to atomic structure, using his analysis of the spectral lines in the light emitted by hydrogen atoms. Bohr explained the frequencies of these spectral lines by introducing (involving in involving in involving change and mass of the electron and Planck's constant. He postulated that an atom would not emit radiation while in one of its stable states, but would do so when it made a transition between states. The frequency of the emitted radiation would be equal to the difference in energy between states divided by Planck's constant. An atom could not absorb nor emit radiation continuously but could do so only in finite steps called *quantum jumps*.

The same year, Hevesy invented the *tracer principle*, which made it possible to study chemical reactions without disturbing what was being measured. A radioactive *indicator* (tracer) could be given to a living organism or put in a test tube in amounts so small that there was no effect on the process under study. The tracer was exactly the same chemically as the molecule that it was tracing.

Radioactive atoms that emit high-energy particles or photons let us keep *track* of molecules as they take part in chemical reactions. He called these molecules *radioindicators*. Today we call them *radioactive tracers, or radiotracers*. The *tracer principle* became the foundation of a new medical specialty, called *atomic medicine*, later being called *nuclear medicine*, and today called *molecular imaging*.

In 1931, when a British physicist, Paul Dirac, was developing *quantum mechanics*, his equations showed the probable existence of an elementary particle, called a *positron*, having the same mass as an electron but with a positive rather than a negative charge. The existence of positrons was subsequently proven, when they were detected in cosmic radiation by Nobel Prize winner, Carl Anderson.

Quantum mechanics made it possible to explain physical phenomena at the level of the atom, which was not possible using classical physics developed by Newton. If Newtonian principles did apply to the behavior of atoms, electrons would be drawn into an atom's nucleus, rather than follow an orbital path around the nucleus, which is the case. The model of the structure of the atom defied classical electromagnetism, which was found to apply only applied only to the world that we see in everyday life.

The University of California at Berkeley was a major factor in the development of nuclear medicine in the United States. William Donner provided funds for the construction of the *Donner Laboratory* at the University of California in Berkeley in 1942, dedicated to the "application of physics, chemistry and the natural sciences to biology and medicine." In 1948, John Lawrence became Associate Director of the Donner Lab. Over the ensuing years, he and his colleagues used radioiron to label the hemoglobin of red blood cells. He carried out pioneering studies of erythropoietin, the hormone controlling the production of red blood cells. The 184-in. cyclotron was used to produce high-energy protons, deuterons and helium ion beams to treat patients with cancer, and to bombard the pituitary in patients with acromegaly.

In 1970, when Ernest Lawrence retired as director of Donner Laboratory, he was asked by Governor Reagan to become a Regent of the University of California. During his 13-year-tenure, he was instrumental in promoting advanced education in the medical sciences. In 1983, he received the Enrico Fermi award for his "pioneering work and continuing leadership in nuclear medicine."

During the last few years of his life, he kept a table next to his bed, filled with scientific and medical books. When unable to sleep, he would get up to read at any time of day or night. As Lawrence's colleague and later Nobel prize winner, Luis Alvarez, wrote: "Lawrence had developed a new way of doing what came to be called 'big science', and that development stemmed from his ebullient nature plus his scientific insight and his charisma; he was more the natural leader than any man I've met." With the help of Arthur Loomis, Lawrence received a breathtaking $1.15 million from the Rockefeller Foundation to build a 184-in. cyclotron, far bigger than the 7-in. and 30-in. and 60-in. machines that had been built previously. In Loomis's words: "It was obvious from the very beginning, when he (Lawrence) was building (radioactive) isotopes, that it opened up matters for making medical measurements as well as chemical and physical measurements." After spending an enormous amount of time generating the funds, a 184-in. cyclotron was finally on the drawing board, when, on September 1, 1939, Germany invaded Poland.

On November 9, 1939, Ernest Lawrence (Fig. 7.1.) won the Nobel prize in physics for "for the invention and development of the cyclotron and for results obtained with it, especially with regard to artificial radioactive elements" Two of Lawrence's students at Berkeley, Luis Alvarez and Edwin McMillan, subsequently received the Nobel prize.

In the early 1940s, Fermi and Szilard at Columbia University were trying to obtain a chain reaction, based on the discovery of deuterium by another Nobel laureate, Harold Urey. Ed McMillan had discovered Uranium-239. Glenn Seaborg and Emilio Segre showed that another product of uranium bombardment with deuterons was a new element, plutonium-239. They would be among many of Lawrence's disciples to receive the Nobel Prize: McMillan for his discovery in 1940 of neptunium; Seaborg in 1951 for his discovery of plutonium; and Alvarez in 1968 for his work in high-energy physics.

Ten years after the invention of the cyclotron, the nuclear reactor was invented. In Germany in December, 1938, Hahn and Strassman discovered fission: a uranium atom could be split into smaller elements. In December 1942, Enrico Fermi and his colleagues built the first nuclear reactor in Chicago as part of the Manhattan Project. The nuclear reactor was able to provide a far wider source of radioactive elements and compounds at much lower cost than the cyclotron.

Fig. 7.1 ErnestL awrence,i nventoro ft hec yclotroni nt hee arly1 930s.

The public was kept in the dark about the development of the atomic bomb during its two and a half years of development by the Manhattan project. Some secrets had leaked out, but most people had never even heard of "radioactivity," a word that was for decades to incite fear in the minds of people all over the world.

J. Robert Oppenheimer, on November 2, 1945 in a speech to the Association of Los Alamos scientists, said that: "...the real importance of atomic energy does not lie in the weapons that have been made. The real importance lies in all the great benefits, which the various radiations, will bring to mankind."

Radioactive elements, especially carbon-14, were key products of the Manhattan Project, and could be produced in large quantities by the nuclear reactors. They would provide the world with new tools for chemical and biomedical research. Radioactive "tracers" were able to "broadcast" their presence as part of "radiolabeled" molecules as they defined the "the stream of life." Being able to measure the chemical processes in every part of the body of living organisms made it possible for creative biological and medical scientists to provide a whole new approach to biochemistry and medicine. The radionuclides, chiefly carbon-14 and phosphorus-32, led the way.

Ernest Lawrence devoted all his efforts to physics, and appointed his brother, John Lawrence, to be Director of the University's Medical Physics Laboratory. Ernest himself became a consultant to the Institute of Cancer Research at Columbia.

In 1946 Paul Aebersold was placed in charge of the government's radioisotope development at Oak Ridge, Tennessee, where he directed the first peaceful distribution and use of nuclear products. The first announcement of the availability of reactor-produced radioisotopes for public distribution was published in the June 14, 1946 issue of SCIENCE.

Eleven years later Paul was transferred to the Atomic Energy Commission (AEC) Headquarters in Washington where he directed the phenomenal growth in the production and use of radioisotopes. "He was responsible, more than any other individual for the acceptance and application of radioisotopes in this country, and contributed a great deal to their use abroad...He became "Mr. Radioisotope" as the result of his energetic, influential, and successful efforts to introduce the use of radioisotopes into a broad spectrum of scientific and practical applications. He retired as Director of Isotopes Development in the AEC in 1965. With his support of investigators, including Bill Myers at Ohio State University, the radionuclides Cobalt-60, Gold-198, Chromium-51, Iodine-125, Iodine-123, Strontium-87m, and Carbon-11 were introduced into biomedicine.

In the American Journal of Roentgenology, Radium Therapy and Nuclear Medicine Vol LXXV, No 6, June, 1956, Paul wrote: "Great advances have been made in nuclear medicine in the last two decades. The work in the first decade with accelerator-produced radioisotopes laid a firm groundwork for development in the last decade of reactor-produced radioisotopes...On the basis of past developments and the present state of nuclear medicine, we can expect a continuous and steady growth in techniques and uses. It is impossible to anticipate all the future benefits mankind will derive from this new field, nuclear medicine."

In the 1940s radioactive iodine began to be used in the diagnosis and treatment of patients with thyroid disease. An article that appeared on June 4, 1963 in the Wall Street Journal unit described the construction of the cyclotron in the Physics department at Washington University. For the first time, the economics of hospital cyclotrons were examined. The public joined scientists in awe of this new approach to examine the chemistry of the living human body.

Hal Gray, in the Medical Research Council facility at Hammersmith Hospital in London, proposed that high radiation could be used to treat cancer. He established a radiation therapy center, a project greatly aided by Michel TerPogossian of Washington University in St. Louis. On January 28, 1953, Queen Elizabeth inaugurated a new cyclotron, and, on June 29, 1953, the first beam was extracted. In October, 1953, the first patients were treated with the linear accelerator.

On June 29, 1956, studies were begun with the radioactive gas, oxygen-15. I was a fellow at Hammersmith Hospital in 1957, where I learned of the work being carried out and planned for the future. I decided then to have a career in nuclear medicine. My work at Hammersmith Hospital consisted of performing studies of the thyroid in patients with iodine-132, obtained by distillation from tellurium-132. Physicist John Mallard told me to look into the newly developed "radionuclide generators" being produced by Stang and Richards at Brookhaven National Laboratory after I returned to Hopkins as Chief Medical Resident in June, 1958.

During World War II, Enrico Fermi and his colleagues at the University of Chicago invented the nuclear reactor. Neutrons could bombard different chemicals to create radioactive tracers more easily and cheaply than was possible with a cyclotron. Therefore, the cyclotron was put on a back burner in biomedical research until the 1970s, when there was a resurgence of interest in carbon-11, oxygen-15, fluorine-18 and nitrogren-13, elements of enormous biological importance that could only be produced in a cyclotron.

In the 1940s, nuclear reactors produced large quantities of carbon-14, tritium, phosphorus-32, and other radionuclides. In 1945, the US government made a key decision: radionuclides produced in US government facilities would be made available for civilian use by qualified physicians and scientists all over the world. The government established training programs in the effective use of radiotracers in biology, biochemistry and medicine. These historic decisions led to the development of modern biochemistry to join anatomy and physiology as the foundation of scientific medicine.

In June 1946, President Truman signed an executive order that made reactor-produced iodine-131 and other radionuclides available from Oak Ridge National Laboratory in Tennessee. On August 2, 1946, the first shipment of carbon-14 was made to Martin Kamen at the Lawrence Laboratory in Berkeley, California. The shipment was kept secret because Kamen was falsely suspected by many of the public to be a communist. The first announced shipment to a medical facility was to the Bermard Free Skin and Cancer Hospital at Washington University in St. Louis.

On December 7, 1946, New York internist Sam Seidlin and his colleagues made the exciting announcement that radioiodine could not just ameliorate but could cure metastatic cancer of the thyroid. Marshall Brucer at Oak Ridge has written that, within days every Congressman heard the news from his constituents. On Janaury 1, 1947, the U S Atomic Energy Commission (AEC) took over the distribution of radioisotopes from the super secret Manhattan District Project of World War II that had developed the atomic bomb. Brucer believed that Seidlin's article was the most important ever published in the history of nuclear medicine.

By the end of 1947, the Atomic Energy Commission had made 2,191 shipments of radioactive isotopes, chiefly carbon-14, to 236 institutions throughout the world. The AEC formed an Advisory Committee on Isotopes, with a subcommittee on their "Use in

Humans," often referred to as the "Subhuman Committee." There were 5,298 shipments in 1949, and the AEC, FDA and NIH began to compete over who should control "isotopes." The AEC charged the users 20% of the cost of production of the radioisotopes. All isotopes came from Oak Ridge, about half of the users having been trained in a 1 month course at the Oak Ridge Institute for Nuclear Studies (ORINS).

In 1947, Robert Hofstadter invented the sodium iodide thallium-activated scintillation crystal, which made possible gamma energy spectrometry, pioneered by P.R.Bell and Craig Harris at ORINS. Spectrometry greatly improved the efficiency of detection of radioactive tracers labeled with multiple radionuclides. Crystal detectors were more sensitive in the measurement of radioactivity both in vivo and in vitro.

The longer half-lives of radionuclides greatly facilitated their availability, use and cost, compared to cyclotron-produced radionuclides. Reactor-produced carbon-14 and tritium emit beta particles. Their emissions have a very short range in tissue, so that they cannot be used with radiation detectors outside of the body to examine regional biochemistry in living human beings or experimental animals. A single photon-emitting radionuclide, technetium-99m, which could be readily obtained as a decay product of molybdenum-99m, was key in the growth of nuclear medicine because of its availability, 6-h half-life, type of decay (isomeric transition), and the optimum energy of its 140 keV photon emissions. Radionuclide generators, called *cows*, provide short-lived radionuclides far away from the site of production of molybdenum. Examples of positron-emitting radionuclides available from generators include: ^{52}Fe/Mn-52m for production of manganese, and the ^{122}Xe/^{122}I generator. Three decades later, physicians and scientists developed a new interest in cyclotrons in order to examine regional biochemistry using the important radionuclides: carbon-11, oxygen-15, nitrogen-13, and flourine-18.

The field of nuclear medicine was given a tremendous boost on December 8, 1953, when President Dwight Eisenhower gave a speech before the 470th plenary meeting of the United Nations in New York, in which he said:

"The more important responsibility of an atomic energy agency would be to devise methods whereby fissionable material would be allocated to serve the peaceful pursuits of mankind... in agriculture, medicine and other peaceful activities. A special purpose would be to provide abundant electrical energy in the power-starved areas of the world.

"Thus the contributing Powers would be dedicating some of their strength to serve the needs rather than the fears of mankind.

"I would be prepared to submit to the Congress of the United States, and with every expectation of approval, any such plan that would, first, encourage world-wide investigation into the most effective peacetime uses of fissionable material, and with the certainty that the investigators had all the material needed for the conducting of all experiments that were appropriate; second, begin to diminish the potential destructive power of the world's atomic stockpiles; third, allow all peoples of all nations to see that, in this enlightened age, the great Powers of the earth, both of the East and of the West, are interested in human aspirations first rather than in building up the armaments of war; fourth, open up a new channel for peaceful discussion and initiative at least a new approach to the many difficult problems that must be solved in both private and public conversations if the world is to shake off the inertia imposed by fear and is to make positive progress towards peace...

"To the making of these fateful decisions, the United States pledges before you, and therefore before the world, its determination to help solve the fearful atomic dilemma – to devote its entire heart and mind to finding the way by which the miraculous inventiveness of man shall not be dedicated to his death, but consecrated to his life."

Eisenhower had three goals (1) to work with the Soviet Union to transform military to peaceful uses of atomic energy; (2) to negotiate nonproliferation agreements with the Soviet Union; and (3) to involve nations throughout the world, large and small, in peaceful efforts to develop atomic energy to peaceful, rather than military, uses. He said: "It is not enough to take this weapon out of the hands of soldiers. It must be put in the hands of those who know how…to adapt it to the arts of peace…This greatest of destructive forces can be developed into a great boon for the benefit of all mankind…if the entire body of the world's scientists and engineers had adequate amounts of fissionable material with which to test and develop their ideas, this capability would be rapidly transformed into universal, efficient and economic usage." Eisenhower resisted the arguments of the American military to keep the knowledge of nuclear science secret, and created an "open" policy with respect to nuclear energy.

> Eisenhower's support of an "open" policy led to the establishment of the International Atomic Energy Agency (IAEA) in 1957. The same year that Eisenhower gave his speech (1953), an armistice was signed ending the Korean War; the structure of DNA was published by Watson and Crick; Edmund Hilary and Tensing reached the summit of Mount Everest; and the Society of Nuclear Medicine was founded in the United States.

Those scientists who produced the atomic bombs in the United States during World War II predicted that diffusion around the world of their secret knowledge was inevitable. In March, 1963, President John Kennedy said: "I am haunted by the feeling that by 1970, there may be ten nuclear powers instead of four (US, the Soviet Union, Britain, and France), and by 1975, 15 or 20." Nuclear weapons were developed by the US in 1945: the Soviet Union in 1949; France in 1960; China in 1964; Israel in 1967; India in 1974; South Africa in 1982; and Pakistan in 1998. The arm's race today is not one of superpowers but of regional powers. Hundreds of countries have to ability to process uranium. The International Atomic Energy Agency (IAEA) has reported that today 49 nations know how to make nuclear weapons. The IAEA has funded 14 programs in Iran, a country that has an active nuclear weapons development policy. Iran continues to build facilities to process, gasify and enrich uranium.

Threatened by nuclear weapons, some nations believe that the best protection is to develop their own nuclear weapons as a deterrent. Others believe that there should be a multinational fuel bank to produce nuclear fuel for the generation of electricity. Warren Buffett has pledged $50 million toward this goal.

When he assumed the Presidency from Dwight Eisenhower, President John Kennedy said: "The world was not meant to be a prison in which man awaits his execution." Addressing the U. N. General Assembly on June 9, 1988, Indian Prime Minister, Rajiv Gandhi, said: "Nuclear war will not mean the death of a hundred million people. Or even a thousand million. It will mean the extinction of four thousand million: the end of life as we know it on earth."

President Ronald Reagan called for the abolishment of "all nuclear weapons," which he called "totally irrational, totally inhumane, good for nothing but killing, possibly destruction of life on earth and civilization. The extinction of the human species is no longer beyond the possible."

During the Cold War between the United States and the Soviet Union, the deterrent effect of nuclear weapons helped maintain their security, because nuclear weapons threatened "mutual assured destruction (MAD)." Deterrence is no longer the principal means of eliminating the threat of nuclear weapons, because the worldwide threat of terrorism involves individual actions rather than governments. Nuclear weapons now make it conceivable that, for the first time in history, the entire planet earth could be destroyed in a matter of days. We cannot ignore the possibility of the destruction of the world, as we know it.

By 1971, Glenn T. Seaborg was able to write: "Perhaps no field of the life sciences typifies the scientific revolution more than the field of nuclear medicine." In the late 1970s, cyclotrons began to move back into nuclear medicine. By 1991, the cover of the Journal of Nuclear Medicine proclaimed: "Clinical PET: Its Time Has Come."

On October 20, 2006, the ninth World Congress of the World Federation of Nuclear Medicine and Biology was held in Seoul, Korea. On October 8, 2006, North Korea had carried out a nuclear explosion in an underground test. The United Nations Security Council denounced the Korean test. China spoke for the first time of the need for appropriate punitive action. The contrast between the high degree of development over the past half century in South Korea compared to the mass starvation in North Korea illustrates the trade-off between the peaceful and military uses of atomic energy, emphasized in the speech by President Eisenhower in 1953.

On February 10, 1934 Frederick Joliot and Irene Curie published the results of bombarding aluminum foil with neutrons. They discovered that almost any element could be made radioactive. When Ernest Lawrence learned of their work, he realized that he had been producing artificial radioactivity (Fig. 7.2.) in the routine operation of the cyclotron

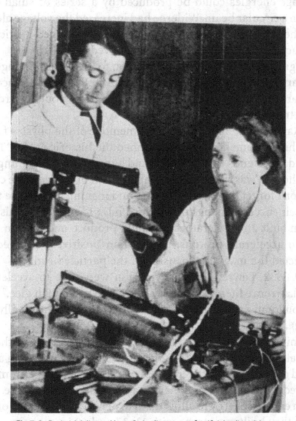

Fig. 7. 2 Frederick Joliot and Irene Curie, discoverers of artificial radioactivity.

that he and his colleagues had invented in 1931. He was not the first to discover artificial radioactivity even though his cyclotron was producing artificial radioactivity in its component parts whenever it was running. The electrical power supply to his radiation detectors was linked to the power supply of the cyclotron. When the cyclotron was shut down, the radiation detectors were automatically turned off as well. Too bad!

Cyclotron-produced oxygen-15 and carbon-11 are the most fundamental elements in living systems. In the 1930s, chemist Martin Kamen, working with Ernest Lawrence, discovered that the oxygen produced by the process of photosynthesis in plants came from water, not from carbon dioxide, as had been previously thought.

Ernest Lawrence invented the first cyclotron with the help of his graduate student, M. Stanley Livingston. Up until that time, physicists could only produce neutron beams by exposing the element beryllium to the alpha particles emitted by naturally occurring radium or polonium.

The cyclotron, produces phosphorus-30, nitrogen-13, carbon-11 and oxygen-15. These radionuclides, plus fluorine-18 (an analogue of hydrogen), became the most important positron-emitting radionuclides in biomedicine. The cyclotron made possible the routine production of these positron-emitting radionuclides.

Two years before Ernest Lawrence went to Berkeley, he was browsing in the library at Yale, and chanced upon an article by a German physicist, R. Wideroe, who showed that it was possible to produce a voltage of 25,000–50,000 V. Lawrence saw that he could use this high energy to bombard atomic nuclei with positively charged hydrogen ions. Higher and higher voltage energies could be produced by a series of small voltage increases properly timed in a circular beam. Wideroe had increased the voltage along two cylinders lined up along the same axis. Lawrence put electrically charged particles in a magnetic field that could accelerate the particles as they moved in a circle.

Lawrence assigned the building of the cyclotron to M. Stanley Livingston, his graduate student. In January 1931, Livingston was successful, and was able to accelerate hydrogen ions and bombard elements. The high energy particles could overcome the nuclear charge and penetrate the atomic nucleus.

A colleague, Raymond T. Birge, who was a member of the physics faculty at Berkeley from 1918 until his retirement in 1955, described the historic event: "The next morning Dr. Lawrence told his friends that he had found a method for obtaining particles of very high energy… The idea was surprisingly simple and in principle quite correct – everyone admitted, but 'Don't forget that having an idea and making it work are two very different things.'" With their success, Livingston got his Ph.D.; Lawrence got his Nobel Prize.

Lawrence then built an accelerator that could produce one million volts by using an 11-in. magnet to accelerate protons, rather than positively charged hydrogen ions. Livingston perfected the magnetic focusing of the particle beams.

On January 9, 1932, Lawrence and Livingston were able to accelerate protons to a 1-MeV energy. "Lawrence literally danced around the room with glee." The two partners then set out to build an even bigger "cyclotron," which they originally called a *circular magnetic resonance accelerator*.

The new machine was completed by the summer of 1932 and by the end of the year, they had accelerated hydrogen ions to 5 MeV, and began bombarding all sorts of targets with protons and deuterons. The latter was better able to enter the atomic nucleus. Accelerated by hundreds of rotations in the alternating magnetic field, the charged particles developed a high enough energy to allow them to penetrate the nuclei of the target atoms. Subsequently by they built cyclotrons with 11, 37 and 60 in. magnets.

As the larger cyclotrons progressively increased in size, the Berkeley group could produce and identify large numbers of artificial radioisotopes, several of which led to Nobel prizes for chemists Emilio Segre and Glenn Seaborg, the discoverer of technetium-99m and plutonium 239. A 184-in. cyclotron bombarded with protons, deuterons and helium ion beams, and began to be used to treat cancer patients and to irradiate the pituitary of patients with acromegaly. Among the early biological studies at Berkeley in the 1930s, prior to World War II, was the clarification of the process of photosynthesis in plants with carbon-11 dioxide, carbon monoxide in human beings and phosphorus-32 in patients with hematological diseases.

Carbon is the most prevalent element in living cells. Carbon defines organic chemistry. Some predict that carbon-11 will someday become the most important radionuclide in nuclear medicine, although fluorine-18 and technetium-99m hold that position today.

Nobel laureate, Glenn Seaborg, discoverer of nine elements, including technetium-99m, wrote: "If we imagine a carbon atom enlarged to the size of a football stadium, the electrons in the outer shell would be like flies flying around the outside of the stadium, and the nucleus would be the size of a golf ball in its center…"

> Carbon became the most common element during evolution, because of its chemical properties. It can donate four electrons to other atoms, or accept four electrons, just like a halogen. Melvin Calvin, a Nobel laureate in chemistry, has written: "Carbon is a constituent of 500,000 known chemical compounds."
>
> Molecular chains of carbon and hydrogen atoms in living organisms extend from a few atoms to long chains, including polymers, such as DNA. Many are enormously reactive in chemical processes because they contain double bonds.

In 1962, the 60-in. cyclotron, called the *medical cyclotron*, was transferred in 1962 from Lawrence's labs to the Crocker Nuclear Laboratory of the University of California at Davis, where it was used primarily for the production of iodine-123.

Ernest's brother, John Lawrence, who received his MD at Harvard in 1930, joined the Radiation Laboratory at Berkeley as a Research Associate in 1936. In February 1936 with his brother, Ernest, as co-author, he published an article on the biological action of neutron rays in (Lawrence and Lawrence, 1936). On Christmas Eve of that year, after successful results in treating leukemia in mice, John carried out the first radiopharmaceutical therapy when he administered a radiophosphorus solution to a 28-year old woman with leukemia. Subsequently this treatment was used to treat patients with polycythemia vera. In 1942, he and his colleagues published their results in Radiology (MLawerence, 1942).

The same year, Paul Aebersold, another graduate student of Ernest Lawrence (Fig. 7.3.), became involved in the worldwide distribution of radionuclides by the US Atomic Energy Commission. He published: *The Cyclotron; A Nuclear Transformer* (Aebersold, 1942). No one did more to make radionuclides available to the biomedical community than Paul.

In 1948, John Lawrence became Associate Director of the Donner Lab. Over the ensuing years, he and his colleagues pioneered the use of radioiron to label the hemoglobin of red blood cells. His colleagues also carried out pioneering studies of *erythropoietin*, the hormone controlling the production of red blood cells.

A cyclotron can produce all of the most important radioactive elements needed for the study of living systems: radioactive oxygen, carbon, nitrogen, and fluorine (a substitute for hydrogen). Carbon defines organic chemistry. The first cyclotron specifically for

Fig. 7.3 Paul Aebersold and John Lawrence. Paul was the first successful promoter of radioisotopes in biomedicine by the Atomic Energy Commission. John was the first to use P-32 in the treatment of leukemia.

biomedical research was built in Cambridge, Massachusetts by physicist, Robley Evans, in November 1940. It was able to produce radionuclides not previously available, and gave birth to the field that would later become *nuclear medicine*.

The first use of positron radiation for examining the human brain imaging was to detect brain tumors. Gordon Brownell of the Physics Research Laboratory (PRL) at Massachusetts General Hospital built a simple positron detection system using two opposed sodium iodide detectors to make possible "coincidence counting of annihilation radiation." When positrons are emitted from a radionuclide, they combine with electrons in the environment, and are annihilated with the resultant production of 511 keV photons that move in opposite directions and can be measure with opposing radiation detectors. Before positrons were used for molecular imaging, scintillation cameras created images from the radiation from tracers that emitted single photons.

Radioisotope scanning made it possible to create images to study one organ after another, including the liver and spleen, organs that could not be seen by conventional x-rays. This new field of nuclear medicine was influenced by dedicated and creative radiologists, including John McAfee at Johns Hopkins, David Kuhl at the University of Pennsylvania, and Merrill Bender at Roswell Park, Buffalo, who were oriented by their training toward anatomy, radiology and surgery, and by internists, such as Joseph Ross and Joseph Kriss, who were oriented more toward physiology, biochemistry and pharmacology.

In 1963, Kuhl and Edwards (1963) developed tomography, years before Houndsfield invented radiographic computed transmission tomography (CT). By 1976 four tomographic imaging systems had been built. Positron emission tomography (PET) was revolutionizing the neurosciences by imaging brain glucose utilization in the human brain after injections of F-18 fluorodeoxyglucose (FDG) and regional cerebral blood flow by oxygen-15 water.

In 1970 J. Adamson showed that modification of the 2-position on the glucose molecule resulted in the persistence of glucose in the brain. Louis Sokoloff, a colleague of Seymour Kety, carried out studies of the rat brain with carbon-14 deoxyglucose. (Carbon-14 does not emit photons, so cannot be used for scanning.) After the successful studies with carbon-14 deoxyglucose, scientists at Brookhaven National Laboratory, the University of Pennsylvania and the National Institutes of Health, developed 18-FDG, a radiotracer that emits positrons that are then converted to photons, making possible to image the regional chemistry of the living human brain. This makes it possible to visualize and quantify the rate of accumulation of glucose or other positron-emitting tracers in different regions of the brain.

In August, 1976, Reivich, Wolf, Kuhl and their colleagues carried out the first Fluorine-18 deoxyglucose (FDG) study of the brain of Dr. Abass Alavi at the University of Pennsylvania (Alavi, 1981). The brain requires a constant supply of energy, most of which comes from glucose and oxygen. The human brain constitutes only 2% of total body mass, but receives 15% of the cardiac output. Its oxygen consumption is approximately 20% of that of the total body.

Hexokinase, which is in the cytoplasm of neurons and glial cells, is a key enzyme in the utilization of glucose as an energy source. After active transport of FDG into cells, hexokinase catalyzes the phosphorylation of glucose to glucose-6 phosphate.

F-18 fluorodeoxyglucose (FDG) is transported into cells by the action of the protein transporters, Glut-1 and Glut-4. FDG-6 phosphate cannot be further metabolized because it is negatively charged, and remains within cells. Glut-1 levels are increased by hypoxia.

Louis Sokoloff and his colleagues at the NIH identified *increased* FDG accumulation at the site of origin of focal epileptic seizures (Sokoloff, 1979) when no structural abnormalities could be seen with computed tomography (CT). Regions of *decreased* glucose utilization in the brain were found in patients with stroke, senile dementia of the Alzheimer type (SDAT), and Huntington's disease (HD). These were often found before structural changes could be seen.

The early clinical studies with PET/FDG were in oncology, where they had a tremendous impact. In late 1930s in Germany, Otto Warburg postulated that malignant tumors are similar to archeobacteria, using glucose for energy. Glucose is the most primitive source of energy in living systems. Anaerobic glycolytic pathway existed in archeobacteria before there was oxygen on earth. Hexokinase activity in the cytoplasm of all nucleated cells increases when cells are damaged.

If cancerous lesions are not growing rapidly, they do not metabolize glucose avidly, and are often not visible by PET/FDG imaging. These include prostate cancer, bronchioalveolar lung cancer, early stages of breast cancer, and renal cancer. They do accumulate C-11 acetate, which reflects oxidative phosphorylation (Krebs cycle)), and C-11 or F-18 labeled choline, a constituent of biological membranes.

Heart muscle consumes glucose avidly during exercise. Fatty acids supply energy to the heart at rest. FDG accumulation is increased in hypoxic infectious processes, such as tuberculous lesions.

Structural imaging modalities, CT and MRI, are complementary to PET imaging. Structural imaging looks at the terrain; PET takes samples of the underlying land. Molecular imaging can examine the chemical processes going on in neurons, axons, dendrites and synapses, and correlate the findings with thoughts, moods, intentions, language, and behavior, the software of the brain.

Examples of PET radiotracers available in institutions with PET facilities include:

1. C-11methyl spiperone and C-11raclopride for dopamine receptors
2. C-11 MDL 100,907 for serotonin receptors
3. C-11 carfentanil for opiate μ receptors
4. C-11 McN5652 and ^{11}C DASB for serotonin transporter sites
5. C-11 WIN-35,428 for dopamine transporter sites
6. C-11 thymidine and ^{11}C FMAU for DNA proliferation
7. O-15 water for cerebral blood flow measurements
8. F-18 fluoro-2-deoxy-D-glucose (FDG) for glucose metabolism

Radioligands are available for the study of opiate receptor subtypes, nicotinic (^{18}F) epibatidine analogs) and muscarinic acetylcholine, ([^{11}C] dexetimide) receptors, benzodiazepine receptors, histamine receptors, and angiotensin receptors.

In the early days of radioisotope scanning, scans were always superimposed over radiographs taken at the same time in the nuclear medicine department. Fiducial markers were used to superimpose anatomy and radionuclide distributions accurately.

The combining of positron emission tomography (PET) with computed tomography (CT) began in 1998 and grew rapidly. By 2004 nearly 1,000 units had been installed worldwide. In the United States, by 2006, ninety-five percent of all PET purchases were PET/CT systems. In addition to being able to look at structure, function and regional biochemistry, the CT data can be used to correct the PET images for attenuation of the 511 kev photons as they pass through the body. This decreases the time required to perform the PET imaging procedure by about 30%.

Accurate fusion of PET and CT images requires correction for the motion of the heart and diaphragm. Fusion of the PET/CT images by hardware or software is both more sensitive and specific in detection of abnormalities than looking at the two sets of images side by side.

7.1. Molecular Imaging

Over the past decades, rapid progress has been made in developing imaging technologies, including functional magnetic resonance imaging (fMRI), diffuse optical tomography (DOT), also called Near Infrared Spectroscopy (NIRS), and magnetoencephalography (MEG). Regions of neuronal activation are detected during mental activities, subjective or in response to external stimuli.

In Molecular Imaging (MI) with radiotracers, molecules labeled with radioactive atoms, which undergo the process of radioactive decay are localized and quantified to produce what are called "molecular images." According to the "tracer principle," these molecules are often identical to a natural substance within the body, such as glucose. The distribution and movement of the "tracer" is imaged within the body with a positron emission tomography (PET) or single photon emission tomography (SPECT) scanner. Radioactive tracer molecules are called "*radiopharmaceuticals*," despite the fact that they have no pharmacological effect when used in diagnostic studies. This is an important characteristic distinguishing them from nonradioactive drugs that are given to produce a pharmacologic effect.

When a radioactive atom emits positrons, the latter are immediately transformed to 511 kV gamma ray photons as a result of the positron interacting with electrons in the environment. These gamma rays travel to the outside of the body where they are measured by radiation detectors in a ring surrounding the body. The emitted photons from the radionuclides that label radiopharmaceuticals make it possible to examine the molecular processes under study in specific regions of the body. The PET scanner contains a computer that portrays the distribution of radioactivity within the body in the form of images that can be quantified PET images reveal regional molecular processes, unlike conventional x-rays, *computed tomography*, or *magnetic resonance imaging*, which reveal body structures.

Other companies, including Picker and Nuclear Chicago focused on imaging devices to detect and image single photon emitting radiotracers. Thus, molecular imaging includes single photon emission tomography (SPECT) as well as positron emission tomography (PET)

7.2. Johns Hopkins Medicine

In 1957, I left the National Institutes of Health in Bethesda, Maryland, to begin a 1-year research fellowship at Hammersmith Hospital. When I saw the exciting work being done with cyclotron-produced radionuclides at Hammersmith, I decided that my goal in life would be to work in the field that was then called "atomic medicine." Later, its name was changed to "nuclear medicine," and today the field is called "molecular imaging." Its basic principle remained the use of radioactive tracers to measure biochemical processes in the living human being. The use of radioactive tracers in biomedical science makes it possible to measure "regional physiology" in different parts of the body, including the brain. In the 1960s the first studies were made of the blood flow to different parts of the lung. "Rectilinear scanning" was able to reveal defects to different parts of the lung caused by clots passing from the legs and causing "pulmonary embolism." Regions of abnormal ventilation could be detected in patients with obstructive airways disease, including chronic bronchitis and emphysema.

Inventions are made to solve recognized problems. At Johns Hopkins (Fig. 7.4.) in the early 1960s there were attempts by Dr. David Sabiston and others to remove massive, life-threatening pulmonary emboli. Newly invented cardiopulmonary bypass procedures made the operation, called embolectomy, possible. Eventually, surgeons found that the

Fig. 7.4 Rectilinears cannerb uilta tJ ohnsH opkinsi n1 959.

stress of the embolectomy was too great, and a preferable treatment was the use of a new drug, called Urokinase to dissolve the clots in the pulmonary arteries.

Perfusion lung scanning was performed with a rectilinear scanner, built at Johns Hopkins and modeled on the scanner invented by Ben Cassen.

We were able to show that the drug treatment facilitated the recovery of blood flow to the involved regions of the lung, and speeded the recovery of patients with acute pulmonary embolism. The Urokinase study was the first large scale, multi-institutional clinical trial. Lung scanning in this clinical trial was the first "surrogate" marker, the use of an objective measure of the clinical response of patients in clinical trials in the process of drug design and development.

An advantage of quantitative imaging of regional physiological or molecular processes is to help select specific patients for clinical trials. Molecular characterization is better than giving the patient a nonspecific diagnostic label. Patients diagnosed as having Alzheimer's disease are often depressed. Patients diagnosed as having bipolar disorder can have trouble with their memory. These diagnoses are man-made categories that are pragmatic abstractions of disease manifestations in very complex patients. Simplifications are possible at times. Dopamine is involved in movement and emotions; serotonin is involved in mood; acetylcholine is involved in intelligence.

The pharmaceutical industry uses PET studies of receptor occupancy to determine the proper doses of new radiopharmaceuticals in Phase I and II clinical trials. Receptor occupancy is assessed in those receptors upon which the new drug acts. An early example was the use of ^{11}C-carfentanil to measure the duration of blockade of opiate receptors by the drug nalmifene, compared to naloxone.

7.3. Society of Nuclear Medicine

In 1950, the Food and Drug Administration (FDA) officially recognized iodine-131 as the first "radioactive new drug." By that year, 3,250 publications had been about the use of radionuclides in medicine. The time had come for a professional organization to be formed to bring together persons interested in "atomic medicine." Robert R. Newell of

Stanford University changed its name to "nuclear medicine."

On Tuesday, January 19, 1954, four internists, four radiologists, three physicists and a cardiac physiologist met in the Davenport Hotel in Spokane, Washington to create what became the Society of Nuclear Medicine. In 1971, the American Board of Medical Specialties officially recognized nuclear medicine as a medical specialty in the United States. The specialty involves the imaging and quantification of molecular and physiological processes within the body of human beings and experimental animals. Radiopharmaceuticals are "imaged" as they travel from one part of the body to another. The initial use in the diagnosis of thyroid disorders was soon extended to the detection of abnormalities of the lungs, heart, kidneys, brain and other organs of the body.

7.4. Hal Anger

From July 1943 to December 1945, Hal helped develop radar jamming, which became crucial in the air war in Europe during World War II. He also worked on anti-jamming methods to counter the jamming by the Germans at that time. In 1948, Hal went to work at the Donner Laboratory, which was part of the Lawrence Radiation Laboratory, named for Nobel laureate, Ernest Lawrence, inventor of the cyclotron in the early 1930s. John Lawrence, brother of Ernest, was head of the Donner Lab, but Hal's immediate boss was the late Cornelius (Toby) Tobias, one of the many Hungarian refugees who came to the US in 1939, and contributed so much to radiation sciences in America. Tobias was a founding member of the Donner Laboratory, who pioneered the study of the biological effects of cosmic rays. He worked with Luis Alvarez and Emilio Segre, two other Nobel laureates in the Lawrence Radiation Laboratory. Tobias became famous for his work in radiation therapy based on the use of the fifth of Ernest Lawrence's cyclotrons, the 184-in. cyclotron, to which Hal was assigned as an engineer. Tobias's career at the Donner Lab, spanned over 40 years. Among his many accomplishments were the use of carbon-11 in biological studies of "bends" in pilots, and in developing xenon gas as anesthetic.

Among the first projects to which Hal was assigned was the modification of the 184 in. cyclotron so that it could be used for irradiation of pituitary tumors with high energy deuterons. During World War II, the 184-in. cyclotron had been converted to a large-scale spectrograph, or *Calutron,* used to produce the uranium-235 for the first atomic bomb. Subsequently a mass spectrograph was built in Oak Ridge to make large amounts of uranium and plutonium, and the 184-in. cyclotron was re-configured for the biological and medical purposes. Hal also worked briefly with Luis Alvarez on the development of contrast agents for cholecystography.

Hal's greatest contribution was to replace the rectilinear scanners with the "scintillation camera" (Fig. 7.5.) invented in 1957, and perfected in the 1960s and 1970s. The invention of the Anger camera led to the birth of nuclear cardiology, which accounts for more than half of all nuclear medicine imaging studies in the US today.

In 1951, Hal Anger read in the journal, *Nucleonics,* of the invention of the rectilinear scanner by Benedict Cassen at UCLA. He thought: "I can do better than that." He recognized that the mechanical movement of the crystal radiation detector over the patient's body as designed in the Cassen rectilinear scanner was a serious limitation, and set out to develop a "gamma ray camera.

The scintillation camera built by Anger was described in an article entitled "A New Instrument for Mapping Gamma-Ray Emitters" in Biology and Medicine Quarterly Report UCRL-3653, p 36 in January, 1957. This camera used a sodium iodide crystal 4 in. in diameter. Hal had had little contact with Ernest Lawrence, but Tobias brought Lawrence to Hal's lab to see the new invention. Lawrence said: "Gee, we ought to try to get this released (from the Atomic Energy Commission, AEC) to Hal Anger so he can get something out of it." With Tobias's and John Lawrence's help, the AEC released the patent rights to the inventor, and the University of California could have claimed the patent rights, but declined to do so. In November, 1958, patent # 3011057 describing the scintillation camera was granted to Hal Anger. The instrument was exhibited at the 1958 annual meeting of the Society of Medicine and subsequently that year at the meeting of the AMA. As expected, the new invention was warmly received. The next step was its commercial development, an important story in itself.

The first of Hal's major contributions to biochemistry and nuclear medicine had been the invention of the well counter for measurement of radioactivity in liquid samples in 1951 (Rev. Sci. Instr. 22, No. 12, 912–914, 1951). Well counters soon became the most widely used instrument in radiation chemistry.

Fig. 7.5 Hal Anger, exhibiting the first scintillation camera at a meeting of the Society of Nuclear Medicine.

The first gamma camera built by Hal Anger used a pinhole collimator (Nature 170, 200, 1952). Gamma photons emitted from I-131 in a patient with metastatic cancer near the skin were the first images with a pinhole collimator that was placed in front of a 2 x 4 in. thallium-activated sodium iodide crystal 5/6 in thick. Another key publication was: The Gamma-pinhole Camera and Image Amplifier (UCRL-2524, 1954).

Hal enlisted the help of a physician, Alex Gottshalk, who came to the Donner Lab in July, 1962. By 1963, Hal and Alex described the "Localization of Brain Tumors with the Positron Scintillation Camera" (J Nuclear Medicine 4:326–330, 1963). This represented the first clinical use of the positron scintillation camera.

It was an improvement over the instrument described in by Hal in January, 1958 (Rev Sci Inst. 29, 2733, 1958). Alex had Hal move the camera to the Alta Bates Hospital in Berkeley where patients with brain tumors could be imaged. The radionuclide used by Anger and Gottshalk was positron-emitting gallium-68, with a half life of 68 min. This nuclide was obtained by the elution from a germanium–gallium generator system. The parent ^{68}Ge has a half life of 275 days. The new camera had an 11½ in. diameter sodium iodide crystal, which made it possible to examine the entire brain. They studied 25 patients and compared the results with ^{203}Hg Neohydrin images obtained with images obtained with a rectilinear scanner.

Hal also collaborated with another Donner Laboratory physician, Donald Van Dyke, using another positron-emitting radionuclide, iron-52, to image the distribution of iron in the bone marrow of patients with hematological disorders and Paget's disease. In 1964, they presented their results at the Eleventh Annual meeting of the Society of Nuclear Medicine at the Claremont Hotel in Berkeley.

At the same meeting, Gottshalk and Anger presented a talk on "Diagnostic Uses of the Scintillation Camera." They used the single photon camera with ^{203}Hg Neohydrin and ^{131}I Hippuran. The camera made dynamic studies possible.

After 2 years at the Donner Lab, Gottshalk left to go the Argonne National Laboratory in Chicago, where he met a surgeon, Paul Harper, who had become attracted by the field of nuclear medicine. In 1960, Stang and Richards had advertised on the cover of the Brookhaven National Laboratory the availability of technetium-99m, obtainable from a molybdenum-99 generator. Three years later, Paul Harper recognized that technetium-99m had physical characteristics that were ideal for nuclear imaging: The nuclide was meta-stable, decaying to Tc-99 without emitting radioactive particles. This made it possible to safely administer large doses, an extremely important advantage for the scintillation camera. Its 160 kev photon emission was high enough to penetrate to the outside of the human body. Technetium99m was readily available by elution from a Mo-99 generator.

7.5. William G. Myers

A key pioneer in the development of the Anger camera was the late William G. Myers of Ohio State University. As a graduate student at Ohio State, Bill's Ph.D. thesis was on the potential role of the cyclotron in biomedicine as the source of positron-emitting radiotracers. Bill had participated in the atomic bomb testing at Bikini prior to becoming an internist and nuclear physician. He traveled every summer to the Donner Laboratory to direct a course in the use of radioactive tracers in biomedicine.

Bill recognized immediately the importance of the invention of the scintillation camera, and persuaded Professor Charles Doan, head of the Department of Internal Medicine at Ohio State, to place an order for the first commercial version. This camera was to be built by a new company, Nuclear Chicago, under the leadership of John Kuranz, President. The company was founded in 1947 by John Kuranz and others from the nuclear reactor laboratory of Enrico Fermi at the University of Chicago. By the early 1970s, Ohio-Nuclear of Solon, Ohio, as well as Picker Medical and General Electric, were also providing gamma cameras to hospitals.

Images were displayed (Fig. 7.6.) using a process called *photorecording*, using x-ray film to display the regional concentrations of radioactivity within the body, including the brain.

On July 30, 1966, Searle purchased the Nuclear-Chicago company. In 1969, Searle formed a joint venture with the United Kingdom Atomic Energy Authority called Amersham/Searle Corporation. Nuclear-Chicago continued to expand under Searle's management, creating a joint venture with Shimadzu Corp of Kyoto, Japan in 1970. In 1972, a reorganization resulted in the creation of Searle Diagnostics, consisting of three divisions: the Nuclear-Chicago imaging subsidiary of Searle became known as the Searle Radiographics.

The development of minicomputers throughout the 1970s allowed nuclear imaging to progress from film-based to computer-based images, including cinematic displays.

At Searle, R. Jaszczak developed the first single photon emission tomographic (SPECT) camera in 1976. The total gamma camera market was $50 million in 1973 (in 1973 dollars), compared with $1 billion in 2005. By 1973, the major vendors of nuclear imaging equipment were Searle, General Electric (GE), Picker, and Ohio-Nuclear.

Ohio-Nuclear was a major competitor of Nuclear Chicago in the 1960s. In 1971 Technicare, which began as a financial services firm, Boston Capital, acquired Ohio-Nuclear. Technicare originally concentrated on nuclear imaging, its efforts paralleling and in competition with Searle Radiographics. In 1976, Ohio-Nuclear developed a dual-head gamma camera. Technicare developed body scanning with computed tomography (CT) in 1975. In 1976 the company became the market leader in the US. Technicare's CT and nuclear imaging equipment was distributed by Siemens in Europe and Japan in the mid-1970s.

Fig.7. 6 The Image Display And Analysis System (IDA) built at Hopkins in 1968, one of the first minicomputers in nuclear medicine.

The technical challenges of the emerging nuclear imaging technologies required considerable investments for R&D. Technicare was the world leader in MRI technology during the 1970s. Between 1979 and 1986, Technicare lost $260 million. Financial analysts questioned Johnson & Johnson's ability to manage the development of high-tech imaging instruments. The vision of James Burke, President of Johnson and Johnson's wedding of pharmaceuticals with imaging technology was his rationale for the acquisition of Technicare in 1979. In the 1980s, Technicare could not compete with GE and Siemens in the fields of CT and MRI due to their high quality product lines all across the medical imaging field.

By 1986 Burke was convinced that Johnson & Johnson could not make Technicare profitable and it was time to get out of the business. Twenty years later, Siemens, General Electric and Philips added radiopharmaceutical development to their production of imaging devices. General Electric acquired Amersham for $9 billion; Siemens acquired CTI for $1 billion.

Interest in the cyclotron returned in the 1970s when the radionuclide ^{18}F- was used as a label for fluorouracil for detecting tumors. ^{18}F-fluorodeoxyglucose (FDG) was developed in the 1970s for the study of regional metabolism in the brain, and subsequently in "actively metabolizing tissues" such as cancer. ^{18}F-joined ^{131}I- and ^{99m}Tc-as the most important radionuclides in the history of nuclear medicine.

In Oak Ridge at that time, officials of three major radiopharmaceutical companies stated publicly that hospitals would never undertake the chemical elution of a generator in preparing doses for administration to patients. Fortunately, this viewpoint was not accepted by other scientists present at the meeting.

At the same time that new instruments, such as the Anger camera, were being invented, chemists were developing technetium-99m labeled tracers, based on the reduction of sodium pertechnetate to technetium-tin complexes, the latter being a more useful chemical form for labeling molecules. Radionuclide generators in which molybdenum-99 decayed to technetium-99m made radiotracers readily available in nuclear medicine clinics in hospitals. One after the other new tests were invented and soon put into clinical practice.

Nuclear medicine was the first medical specialty to incorporate computers in medical imaging. Dedicated "minicomputers" were produced by the Digital Equipment Corporation and first used at Johns Hopkins in 1968. They facilitated quantitative analysis of lung scans in the Urokinase/pulmonary embolism clinical trial. Computers, coupled to the Anger camera, were next used to measure global and regional function of the heart. Mathematical models provided computer-generated algorithms for reconstruction and analysis of PET images as well as CT and MRI images, which were later fused with PET and SPECT images.

George Hevesy laid the foundation of nuclear medicine, the tracer principle. The emission of photons from radioactive atoms makes it possible to "track" molecules as they participate in chemical processes anywhere in the human body. Radioactive molecules emit photons that penetrate the body and can tell us what they are doing at all times.

In 1923, Hevesy published a paper: "The absorption and translocation of lead by plants" in Biochemical Journal 17, 439, 1923. Measurement of radioactivity is 1,000 times more sensitive than chemical assays. A radioactive atom can be used as an "indicator" of stable atoms of the same element.

Born in Budapest in 1885, he began his first radioisotope studies in plants in 1923. In 1934, he left Berlin for political reasons to go to Copenhagen to work with Niels Bohr. In 1935, Ernest Lawrence sent phosphorus-32 by regular mail from California. Hevesy published more than 400 scientific articles, and won the Nobel prize in 1944. In 1959, he received the Atoms For Peace award by the US Atomic Energy Commission. He died on July 5, 1966 in Freiburg, Germany.

The first use of radioactive tracers to provide diagnostic information in human beings was in 1925. A Boston physician, Herman Blumgart, injected solutions of the naturally occurring radioactive gas, radon, into the arm veins of patients in order to measure the "velocity of the circulation." He defined it as the time it took for the intravenously injected radon to reach the opposite arm.

In 1940, Hamilton and associates in Los Angeles and Hertz and his associates in Boston were the first to measure the uptake of radioactive iodine-128 by the thyroid. This made it possible to diagnose patients with increased or decreased thyroidal function. The names, hyper- and hypothyroidism, reflected the functional basis of the diagnosis. The use of radioiodine to define thyroid disease led to the search for other applications of the *radioactive tracer principle*.

In 1946, George Moore, a neurosurgeon at the University of Minnesota, used a radio-labeled dye, I^{131} iodofluorescein, to localize brain tumors with a Geiger-Muller detector during surgery.

In 1957, at Hammersmith Hospital in London, I used iodine-132, another radioisotope of iodine, to study patients with thyroid diseases. The iodine-132 was obtained by distilling it from its parent radionuclide, tellurium-132. John Mallard, a physicist at Hammersmith, advised me to look into the newly developed "radionuclide generators" invented by Stang and Richards at Brookhaven National Laboratory when I returned to Hopkins in June, 1958. Generators (called "cows") involved the chemical elution of a short-lived "daughter" radionuclide from a resin column containing a longer-lived "parent" radionuclide.

The same year (1958), TerPogossian and his colleagues at Washington University began studies of the brain with oxygen-15, a radionuclide produced in the cyclotron installed in the physics department in the early 1940s.

In 1973, James Robertson was the first to use a ring of radiation detectors surrounding the patient's head. This made it possible to create images of the distribution of the tracer within the brain. A commercial version of this instrument was developed a decade later by the Cyclotron Corporation.

During the later 1970s, TerPogossian and his colleagues built a series of positron emission tomography (PET) scanners that were later developed commercially by ORTEC, Inc, a company that later was called CTI, Inc, and in 2006 was acquired by Siemens AG.

TerPogossian's cyclotron had been in the Physics Department. The first cyclotron in a medical facility and used exclusively for biomedical research was built at Hammersmith Hospital in London in 1955, under the auspices of the Medical Research Council. A similar cyclotron was subsequently built at Massachusetts General Hospital in Boston. Using a medically dedicated cyclotron that was built later at Washington University, Raichle and colleagues showed that vision and other mental activity resulted in an increased oxygen metabolism in the involved regions of the brain.

The studies of cerebral blood flow with radioactive tracers were an advance over the prior studies carried out in 1948 by Kety and Schmidt (1948), using the nonradioactive gas, nitrous oxide. They showed that blood flow per unit volume of brain declined with advancing age, the greatest decline occurring in childhood. Decreases in brain blood flow were greatest if the older patients had cerebrovascular disease.

Subsequent studies with the radioactive gases, krypton-85 and xenon-133, showed that the decrease in cerebral blood flow with age was in gray, not white matter, and involved the cortex, basal ganglia, and cerebellum (Frackowiak et al., 1980).

Michael Phelps, Chugani and colleagues at the University of California at Los Angeles, carried out PET studies of brain metabolism with fluorine-18 deoxyglucose (FDG) in growing children. Phelps later told writer Joseph Dumit:

"...I inject a molecule that contains a source that can report back to me after I inject it into the body. I can watch that molecule as it fused image goes through the blood supply, goes into the brain, then into the tissues of the brain, and then participates in a chemical process. It does so in a way that does not disturb the process."

He continued: "All disease starts with an error in a chemical process. It typically erodes away the chemical reserves, the compensatory capability of the system to accommodate to the abnormalities. Eventually, the person becomes symptomatic, as the process has eroded way the reserves.

"For the first time, we watched the human brain develop from birth to adulthood...at birth, the areas that were functioning in the newborn child were, not surprisingly, the phylogenetically older parts of the brain: the cerebellum, the central structures of the brain the thalamus, and the old motor cortex... as you watch (the child grow) one structure after the other matures,
...in the final period as the child becomes an adult, the frontal cortex activity is prominent...By the age of 2 or 3, the metabolic rate of metabolism of glucose is twice that of an adult, and remains so until the age of ten."

Shortly after birth, most metabolic activity in the brain is in the phylogenetically older parts of the brain: the cerebellum, the thalamus, and the motor cortex. At that stage of development, the baby's activity consists only of suckling, startle reactions, and other simple reflex actions. As the child develops, the increasing complexity of behavior is accompanied by increased metabolic activity of neurons, reflected in the FDG accumulation. With maturation of mental functions, one region of the brain after another becomes metabolically active.

PET makes it possible to measure molecular processes, not just the state of chemical concentrations in the brain. One can measure (1) delivery of the tracer to the brain after intravenous injection; (2) transport of the tracer from the circulating blood into the brain; and (3) the metabolism of the tracer within the brain and (4) the fate of metabolic products of the tracer. In the case of deoxyglucose, the principal metabolic process is the phosphorylation of deooxyglucose catalyzed by the action of the enzyme, hexokinase.

Sokoloff examined the kinetics of carbon-14 deoxglucose metabolism in the brain of animals. In 1982, Shlomo Levy began producing ^{18}F from oxygen-18, using the reaction ^{18}O (p,n) ^{18}F. In 1986, Hamacher synthesized ^{18}F fluoride rather than ^{18}F$_2$ gas. These advances in the production of F-18 led to the important subsequent development of fluorine-18 deoxyglucose (FDG).

The development of fluorine-18 deoxyglucose by Itoh, Wolf and colleagues made it possible to extend the autoradiographic studies of Sokoloff to living human beings. When Sokoloff injected C-14 deoxyglucose into animals with one eye closed, there were vertical columns of increased glucose metabolism extending down from the surface of the cortex, the greatest increase in layer 4.

Radiolabeled glucose is metabolized too rapidly to serve as a tracer for measuring glucose utilization by PET, which is why it is necessary to use an analogue of glucose, 2-deoxy-D-glucose (2-DG). With fluorine-18 deoxyglucose, it was possible to obtain positron emission tomography (PET) in vivid color images that showed glucose utilization throughout the living human brain, at rest and during mental processes, such as seeing and hearing.

Giovanni DeChiro at the NIH used fluorine-18 deoxyglucose (^{18}F-FDG) to detect brain tumors. Glucose utilization was greatest in most malignant tumors. Studies of brain tumors with FDG/PET are now widely used to detect and determine the degree of malignancy, to help select sites within the brain for biopsy, to determine prognosis, and to monitor the response to treatment. FDG/PET is also used to distinguish recurrence of a tumor from radiation necrosis.

7.6. Bioenergetics

The high glucose utilization of the brain requires a continuing supply of glucose to provide the energy needed to produce ion gradients, called "action potentials," that carry information throughout the brain along axons and dendrites. Glucose is actively taken up as a result of the action of protein transporters, glut1 and glut3. After reaching the brain, glucose is phosphorylated by the enzyme, hexokinase, to produce glucose-6-phosphate. The rate of phosphorylation is limited by the maximum velocity of the hexokinase reaction. Glucose-6-phosphate is then involved in the synthesis of carbohydrates, glycoproteins, nucleic acids and 5-carbon sugars.

The avid accumulation of fluorine-18 deoxyglucose (F-18 FDG) in active regions of the brain is the result of the tracer not being metabolized, but is fixed at the sites of neuronal activity. Deoxyglucose has kinetic behavior significantly different from that of glucose, so that a *lumped constant* must be used to relate deoxyglucose measurements to those of glucose.

F-18 deoxyglucose PET imaging is widely used in the care of patients with cancer. Malignant tumors avidly accumulate F-18 fluorodeoxyglucose (FDG, while lesser malignant or benign tumors do not. FDG accumulation reflects anerobic metabolism. The enzyme, hexokinase II, is found in archeobacteria that existed long before there was oxygen on earth. Cancer reflects de-differentiation of cells to a more primitive, preaerobic state. Malignant tumors are often hypoxic, the result of the expansion of the tumor beyond its blood supply, as occurs in many malignant tumors. The anerobic tumors can be studied with the C-11 acetate, which reflects aerobic metabolism, or by radiolabeled amino acids, such as F-18 tyrosine.

8

The Tracer Principle

In living systems, what we call structures are slow processes of long duration. What we call functions are fast processes of short duration. Imaging procedures, such as computerized tomography (CT) and magnetic resonance imaging (MRI), reveal structure; positron emission tomography (PET), single photon emission computed tomography (SPECT) and optical imaging reveal function and underlying chemical processes. Imaging lets us peer into the living body to identify abnormalities and then tailor treatment to address problems. Treatment is designed for a specific patient. Ideally, one would identify patients at high risk of development of disease before symptoms manifest. Monitoring a person's health should begin even before birth, with the fetus in utero, with special attention being paid to the brain.

In 1848, the French neurologist Paul Broca was the first to relate specific body functions to specific regions of the brain. Today, we can relate the insula to disgust; the anterior cingulate to decision-making; the medial orbital prefrontal cortex to reward-seeking; and superior temporal sulcus and inferior frontal cortex to feelings of empathy.

Broca observed that damage to the brain of his patient, Phineas Gage, suddenly brought about a dramatic change in his personality. He was no longer the sober, industrious person he had been. He became a foul-mouthed drunkard, drifter, and total failure. Broca found damage to his left temporal lobe, which is seen in some patients suffering from a stroke or brain tumor. Broca's identification of the left frontal convolution as being involved in speech led to the naming of the motor speech area Broca's convolution.

Focal lesions often involve vision. Patients with temporal lobe lesions at times cannot perceive movement, although they can see stationary objects. Photons entering the eye are converted into electrochemical processes that create action potentials that progress to the occipital lobes, where visual information is integrated with information in the temporal and occipital lobes. The electrical and chemical activities in the brain determine one's thoughts, feelings, goals, plans, and actions, which we call consciousness.

The prefrontal cortex enlarged during the evolutionary development of the human brain. The amygdala and prefrontal cortex are involved in emotions, including anxiety, fear, and anger. Judgment involves the prefrontal cortex as well as the hippocampus, the part of the temporal lobe involved in memory and cognition. With aging, people are less responsive to threats and aggravation, and become more at peace with themselves and the rest of the world. Older people retain the ability to focus attention and resist distractions.

Automatic nervous system functions are still performed effortlessly, and most knowledge and skills developed over decades remain intact. The ability to link people's facial expressions to their different emotions decreases with age.

People vary in their ability to have pleasurable feelings. Selfishness varies among people. Both personal and societal factors can lead to aggression and conflicts. Genetic and cultural values can lead to altruism, confusion, anxiety, good, and evil.

Mental processes begin with perception, which brings about emotions that are modulated by thoughts and ideas, in the context of cultural and societal values, prior experiences, and memories.

Nature and nurture determine behavior. Matthew wrote in his Gospel, "Love your enemies ... Bless them that curse you, do good to them that hate you, and pray for them which despitefully use you, and persecute you." Martin Luther King said, "I'm sick and tired of violence ... I'm tired of hatred. I'm tired of selfishness. I'm tired of evil."

Aggressiveness is strongly related to the autonomic nervous system, related to the emotions called fight or flight. One can ask, What chemical processes were going on in the brains of those persons involved in a race riot that took place in August 1964, in Watts, South Central Los Angeles, that lasted for 6 days, leaving 34 dead, more than a thousand people injured, 4,000 arrested, and hundreds of buildings destroyed.

President Lyndon Johnson expressed the hope, "We shall never achieve a free and prosperous and hopeful society until we have suppressed the fires of hate and we have turned aside from violence—whether that violence comes from the nightriders of the Klan, or the snipers and the looters in the Watts district, neither old wrongs nor new fears can ever justify arson or murder."

Herbert Benson, the founding president of the Mind/Body Medical Institute at the Beth Israel Deaconess Medical Center in Boston, describes what happens to brain chemistry during periods of stress: "Stress comes from any situation, or circumstance that requires behavioral adjustment. Any change, either good or bad, is stressful, and whether it's a positive or negative change, the physiological response is the same. There is a secretion of epinephrine and nor-epinephrine, hormones that change the mental, as well as the physical components of our body. They lead to increased anxiety, increased anger and hostility, increased mild and moderate depression. They contribute to high blood pressure, hypertension, most heart disease, and angina pectoris. Even heart attacks can be influenced by these hormones.

Humans have always been under stress. There as always been famine, pestilence, and interpersonal problems. But what's different about modern-day life is the sheer amount of information and number of circumstances to which we have to adjust. We are eliciting the fight-or-flight response that frequently leads to symptoms."

In the 1960s, people were fascinated by Indian yogis who were able to alter their states of consciousness by meditation. They were able to control bodily functions, including blood pressure. Benson believes that meditation can help treat high blood pressure, chronic pain, insomnia, and many other physical ailments, as well as in management of stress. Meditating lowers the levels of stress hormones, and boosts the alpha waves to levels seen during sleep.

Andrew Newberg and colleagues at the University of Pennsylvania measured regional changes in glucose accumulation in the brain of subjects during meditation (Newberg et al., 2006). ^{18}F-Fluorodeoxyglucose (FDG) PET was used to determine changes in cerebral glucose metabolism during meditation. Immediately after baseline scans, the subjects performed the relaxation technique for 20 minutes. Five minutes after beginning the relaxation procedures, the subjects were injected with 10 mCi of FDG and scanned 30 minutes later. There were significant increases ($P < 0.05$) in the prefrontal cortex bilaterally, the inferior frontal, the medial frontal, the medial occipital lobe, and the superior parietal lobe. There was a significant decrease in the cerebral glucose metabolism in the thalami and hippocampus ($P < 0.05$). The changes were more prominent in the left hemisphere.

8.1. Dementia

Severe memory loss (dementia) is a major health problem being addressed by molecular imaging. PET, PET/CT, SPECT, and SPECT/CT reveal regions of decreased neuronal activity in persons with memory loss. These characteristic findings help in the diagnosis of mild cognitive impairment (MCI), senile dementia of the Alzheimer type (SDAT), and multi-infarct dementia. PET and other molecular imaging techniques can detect regional chemical changes before structural changes can be seen by CT or MRI.

An important topic of current research is detecting amyloid plaques and neurofibrillary tangles. These plaques were first detected by histopathology, and are caused by abnormal metabolism of the amyloid precursor protein (APP) normally found in cell membranes. The amyloid leads to neurofibrillary tangles, which are intracellular aggregates of tau proteins, i.e., hyperphosphorylated forms of normal proteins. Tau proteins lead to degeneration of proteins. Protective mechanisms in the brain, retina, and lens involve αβ crystallin, which consists of a core β sandwich and an unstructured C and N terminus. Many domains of αB crystallin protect against the accumulation of toxic amyloid forming proteins during the process of molecular aging.

A gene on chromosome 19 contains an allele that encodes β amyloid plaque, and it has been proposed as a marker for Alzheimer's disease. Because there are so many (approximately 60) different causes of dementia, a specific test for early Alzheimer's disease would be very helpful. Characterization of a specific biochemical abnormality in a specific patient would help exclude other causes of dementia (Hammoud, 2007).

Disease manifestations in demented patients can be very complex. For example, patients diagnosed as having Alzheimer's disease (SDAT) are often depressed, or they suffer from bipolar disorder. Correct characterization of specific abnormalities would be an important step forward. Movement and emotional disturbances could be linked to dopamine, serotonin to mood, and acetylcholine to intelligence.

When studied with FDG PET, patients with Alzheimer's disease were found to have bilateral temporal-parietal hypometabolism not seen in other demented patients. Glucose metabolic activity is normal in the basal ganglia, thalamus, cerebellum, and sensory cortex. The sensitivity of FDG in detecting SDAT is about 90%, with the specificity also being 90%.

The search for tracers that bind to amyloid plaques followed the finding of characteristic histopathological manifestations. ^{18}F-FDDNP binds to both plaques and neurofibrillary tangles. Other ligands under study are Pittsburgh compound B and AV-45.

MRI makes possible measurement of the volume of different regions of the brain, such as the hippocampus or entorhinal cortex. It makes possible measurement of total brain volume and the rate of brain atrophy as well as assessment of the integrity of white matter. An automated technique in MRI called voxel-based morphometry can detect the loss of gray matter in regions, such as the temporal lobe gyrus, posterior cingulate, and precuneus. These regions are involved in patients with MCI as well as SDAT. Objective measurements with MRI are more useful than the psychological Mini-Mental Status Examination or AD Assessment Cognitive Scale.

Beginning in the late 1980s, more than a dozen major pharmaceuticals companies, including Upjohn, Bristol-Myers Squibb Pharmaceuticals, and Merck Sharp & Dohme, have tried to find drugs that block the production of β-amyloid. Before cell can release β-amyloid, it must be cut loose from the rest of the APP protein by an enzyme. Thwarting this enzyme is the approach of researchers at Athena Neurosciences, which, with backing from the drug giant Eli Lilly, has one of the world's largest programs of research on Alzheimer's disease. Inflammation also contributes to the pathology of Alzheimer's disease. β-Amyloid may provoke an immune response. The development of vaccines against amyloid plaques in SDAT is another approach.

8.2. Biological Psychiatry

Better understanding of the roles of acetylcholine, dopamine, epinephrine, norepinephrine, histamine, and serotonin in the brain brought psychiatry into the mainstream of medicine and led to the development of effective medications.

Today, one third of all prescribed medicines in the United States are given to affect mental function. Drugs affecting the serotoninergic or norepinephrine systems are helpful in treating patients with depression, panic disorder, anxiety, or obsessive-compulsive disorder. Selective serotonin reuptake inhibitors increase the levels of serotonin and norepinephrine in the brain. In patients with schizophrenia, blocking the effects of dopamine diminishes auditory and visual hallucinations as well as paranoia. Fluoxotin (Prozac), fluoxomine, and phenfluoromine dramatically increase serotonin levels. Differences in serotonin synthesis in men and women explains the higher incidence of depression among women.

Bacterial and viral infections, and metabolic illnesses, such as thyroid disease and diabetes, can cause mental dysfunction. So, too, can abused drugs, fever, dehydration, electrolyte imbalances, toxins, or antibiotics. Patients with HIV/AIDS often become depressed or demented. Patients with multiple sclerosis or cerebrovascular disease, brain trauma, or brain tumors often have mental disabilities.

Radioactive tracers that were developed for PET/CT or SPECT/CT are being converted to fluorescent tracers for optical imaging with near infrared light. MRI and CT also are using molecular methods. MRI researchers are developing carbon-based MRI-signaling molecules; optical imaging tracers can measure physiology or biochemistry.

New cadmium/zinc detectors are being developed for SPECT/CT. Dedicated software is being improved for image display and analysis, interpretation, and communication of results.

9

Aggression vs Altruism

Aggressive behavior is related to the monoamine neurotransmitters: serotonin, dopamine, and noradrenaline. The enzymes monoamine oxidase (MAO) A and B play roles in the metabolism of catecholamines in the brain and peripheral tissues. MAOA degrades dopamine, serotonin, and noradrenaline.

Brunner et al. (1993) found that aggressive behavior was associated with a mutation of the gene for the enzyme MAOA (Science 262:578–580, 1993). In men with borderline mental retardation and aggressive tendencies, they found greatly reduced MAO activity. Mutations of the C936T gene were related to criminal/antisocial behavior.

Childhood maltreatment can lead to antisocial behavior that is related to MAOA activity (Huang et al., 2004). Males with a genotype encoding low MAOA enzyme activity were more likely to be convicted of a violent crime than maltreated males with a high-activity MAOA genotype.

Genes responsible for metabolic functions are active in all cells all the time, whereas others are expressed only when cells begin to differentiate and continue to be expressed after differentiation. Promoter genes stimulate gene expression. Hormones, such as thyroid-stimulating hormone are involved in turning genes on or off. Most physiological processes involve multiple genes.

Nobel laureate Eric Kandel, in his book The Search For Memory, examined the changes in neuronal chemistry related to memory and learning. He won the Nobel Prize in 2000 by relating serotonin and other neurotransmitters to learning and memory. Declarative memory stores faces and events in the hippocampus, the principal site for these memories, experiences and thoughts, encoding them for long-term storage. The hippocampus also stores the ability to walk, drive a car, and perform other motor tasks.

Kandel and colleagues carried out their studies in the giant marine snail *Aplysia*, because it was possible to examine individual neurons. They found that memory does not reside in specific neurons but in the synaptic connections among neurons, which are augmented or weakened by past experiences.

Do genes encode altruism? Persons inclined to be altruistic rather than competitive are more likely to survive and reproduce. Thus, altruism is a driving force in the process of evolution. Genes related to altruism increase the likelihood of survival of altruistic societies.

Darwin thought that altruistic moral tendencies increase the probability of survival. In his book *The Descent of Man* (Darwin, 1871), he wrote, "Man is the rival of other men;

he delights in competition, and this leads to ambition which passes too easily into self-ishness." He thought that aggressiveness had no survival value.

Steve Pinker in his article *The Moral Instinct*, quotes Joshua Greene, a cognitive neuroscientist, as saying that evolution caused people to object to the manhandling of persons innocent of committing crimes.

Using functional MRI, cognitive neuroscientist, Jonathan Cohen, identified the medial frontal lobes as being involved in emotions relating one person to another, whereas the dorsolateral frontal lobes were activated by reasoning. An older part of the brain, the anterior cingulate cortex, is activated when there is a conflict between an emotional urge coming from one part of the brain, and a reasoning advisory is coming from another.

Different cultures, as well as biological factors, have played an important role in the evolution of human beings. Knowledge and imagination were possible only when human beings became able to speak.

Jacob Bronowko (1977) wrote, "We use vision to give us information about the world and sound to give us information about other people." John Calvin said, "Where there is no vision, the people perish."

Acquired characteristics are not inherited in a single generation, but it may take thousands of generations. Cultural characteristics evolve, just as innate biological instincts. One's genes affect the choice of a mate. Acquired characteristics, such as the ability to speak, can make a person more likely to produce children, and behavioral characteristics survior.

Children inherit genes associated with cultural values that enhance survival but also genes that result in aggressive and competitive behavior. Instincts for killing others have had survival value over thousands of years. European countries have been engaged in warfare during 47% of those years (Sorokin, 1957).

In his book *On Aggression* (Lorenz, 1966), Konrad Lorenz says that a visitor from another planet would not conclude that human behavior is dictated by reason, intelligence, or moral responsibility. He would think that human emotions overrule reason and cultural forces. Human behavior is primarily the result of instinctive behavior.

Religious people believe that evil started in the Garden of Eden. John Calvin described the large number of people persons who were bound in chains, lashed, and carried about as laughing stocks. Others were outlawed, cruelly tortured, and could only escape by flight.

Watching the harmful instincts of animals can convince people of the instinctive evil in human beings. They are peaceful with their own families and clans, but they can become ferocious toward those outside their family or community. Some genes inhibit aggressive behavior toward family and community members, whereas others encode aggressiveness toward those outside. Altruistic genes prevent self-destruction of a species. In humans, genes encoding moral responsibility and inhibiting aggressiveness evolved by increasing reproductive attractiveness.

To bomb cars, the bomber must overcome instincts that inhibit aggressiveness. But, according to Lorenz, reason can only modify but not eliminate instinctive, emotionally driven aggressive behavior. Genetic and cultural factors are woven together in a complex way to determine how a person acts.

Reason does not determine behavior; it only rationalizes behavior directed toward achieving emotionally driven goals. Someday, the phenotypic expression of genes encoding aggressive and violent behavior may become less prevalent, supplanted by phenotypic expression of genes encoding altruism.

Thus, one can conclude that moral forces, faith, guilt feelings, and obedience to commandments and laws result from both genetic and environmental factors. Molecular medicine makes it possible to examine the phenotypic expression of molecular processes in the brain that interact with these forces. Molecular processes help define one's personality. These can be characterized quantitatively as deviations from normal molecular processes. Disease can be defined in terms of abnormalities in one or more regional biochemical processes.

Today, in the *International Classification of Diseases*, six digits designate diseases: the first three digits describe the location of the abnormality. The next three digits place each disease in one or more categories, each of which is designated by a six-digit number.

Isolated molecular processes cannot give full insight into the relationship of brain chemistry and behavior. A more complete understanding requires the development of models that describe extremely complex systems with hundreds of molecular processes related to traits and behavior.

It is possible to relate brain chemistry and behavior by means of matrix algebra. One can relate specific neurotransmitters and hormones to specific mental functions in vertical axis (columns) and mental processes on the horizontal axis (rows). The entries in the matrix can be used to relate the relative roles of different complex systems of neurotransmitters to different mental functions. Matrices have long been used to solve multiple linear equations, and they have played an important role in the development of computers after World War II.

10

MonitoringMe ntalA ctivity

With positron emission tomography (PET) and functional magnetic resonance imaging (MRI), one can map out the regions of neuronal activity in the brain from birth to death in experimental animals and humans. The encoding of information in the brain is based on networks of activated neurons and specific molecules that contain information encoding features of the environment. Neuronal networks encode features of objects in the external world. The brain also stores "molecules of the mind."

Oxygen-15 water and F-18 deoxyglucose (^{18}F-FDG) studies of global and regional brain activity help direct attention to activated regions of the brain, which can then be the focus for the study of specific neurochemical processes, such as neurotransmission. The patterns of the abnormalities of neuronal activity in the brain help differentiate benign forgetfulness from mild cognitive impairment (MCI), Pick's disease, or Alzheimer's disease (senile dementia of the Alzheimer type).

The type of anesthesia can have a great effect on glucose use by the brain. In monkeys anesthetized with ketamine, ^{18}F-FDG accumulation in the brain was 3.2 times greater than that in the conscious state and about 4.5 times greater in the cerebral cortex (Takashi Itoh, Osaka University, Japan). Under pentobarbital anesthesia, ^{18}F-FDG accumulation in the occipital cortex was lower than when the animals were awake. The different effects of ketamine and pentobarbital are thought to be due to differences in hexokinase activity.

An example of FDG/PET studies of specific behavior is the study of swallowing. The left sensorimotor cortex, cerebellum, thalamus, precuneus, anterior insula, left and right lateral postcentral gyrus, and left and right occipital cortex are activated during swallowing [Harris et al., J. Cereb. Blood Flow Metab. 25(4):520–526, 2005]. Neuronal activity decreased in the right premotor cortex, right and left sensory and motor association cortices, left posterior insula, and left cerebellum.

When normal subjects were shown film clips that provoked the emotions of happiness, fear, or disgust, oxygen-15 water studies of regional brain blood flow showed increased neuronal blood flow in the primary and secondary visual cortex, as well as in limbic regions [Paradiso et al., Am. J. Psychiatry 154(3):384–389, 1997].

Elderly subjects who suffered from dementia improved their communication skills and motor function by participating in a program of mathematical and reading-out-loud exercises (Yuta Kawashima, Tohoku University, Japan). The mental exercise program called Learning Therapy has been published. The Nintendo Co. has developed a training video program developed on the basis of functional MRI studies of regional brain blood flow.

PET with FDG was used to compare smokers and nonsmokers who were shown ciga-rette-related videos designed to induce cigarette craving (Arthur Brody, Brentwood Bio-logical Research Institute, UCLA, CA). They found that glucose use increased in several regions of the brain when the smokers craved cigarettes. The anterior cingulate gyrus, prefrontal cortex, and anterior temporal lobe (regions involved in anxiety and other emotion) became more active when smokers were given cigarette-related cues. The increase in metabolic activity in the prefrontal cortex was greatest when craving was greatest.

Takahashi et al. found increased dopamine release by nicotine in cigarette smokers but not in nonsmokers. Dopamine release enhances the rewarding effect of smoking in those who become dependent. Nicotine activates the ventrotegamentum area that leads to release of dopamine.

In whole body studies with ^{18}F-FDG during exercise, glucose use shifted from abdom-inal organs to working muscles. Activation during running was greatest in the occipital cortex than in the sensorimotor cortices, suggesting that an important function of the brain during running is processing visual information, whereas neuronal activity con-trolling movement itself is automatic and requires less attention and energy (Masatoshi Itoh, Tohoku University, Japan).

In performing complex tasks, regional cerebral blood flow to the anterior cingulate and supplementary motor cortices increased less in schizophrenic patients than in normal persons (Henry Holcomb et al., Am. J. Psychiatry 157:1634–1645, 2000). The normal control subjects also had progressively greater increases in blood flow to the frontal cortex than did schizophrenic patients, who had lower blood flow and slower response times. The failure to increase cingulate and frontal blood flow suggests that patients with schizophrenia cannot effectively activate frontocingulate neuronal systems involved in performing complex tasks.

While performing a memory task, cannabis abusers showed less neuronal activation than control subjects in the middle temporal gyrus but areas of higher and lower activa-tion within the parahippocampal gyrus, together with differences in hippocampal acti-vation (Block et al., 2002).

Methamphetamine abuse is more common than the abuse of cocaine, marijuana and heroin combined, according to a survey by the National Association of Counties. London and colleagues reported corticolimbic abnormalities in methamphetamine-dependent subjects during early drug abstinence (London et al., 2004).

Ancient philosophers, beginning with Aristotle, believed that all knowledge is derived from sense impressions that are stored in the brain and become the basis of thinking. Aristotle wrote, "no one could ever learn or understand anything if he had not the faculty of perception; even when he thinks speculatively, he must have some mental picture with which to think." The mental pictures that we take from sensory impressions are like a painted portrait, "the lasting state of which we describe as memory."

The concept of *consciousness* may someday just fade away from attention, as have other concepts that were once hot topics. Someday, it may be possible to examine the chemistry of isolated individual cells and image their functions as they work together within the living brain.

We constantly perceive the world, think about what we perceive, relate it to memories, and convey our thoughts to others. Thought is linked to speech. Nouns and verbs express thoughts. Not only things, but also concepts, such as truth, justice, courage, and divinity, are thoughts in words, as proposed by Hannah Arendt ("The Life of the Mind" Harcourt Brace Jovanovich, San Diego, 1971). With PET, both thoughts and words can be physically localized in the brain.

We can try to find out what goes on in neurons when we are seeing and/or speaking Images and language involve separate domains in the brain. Pictograms and hieroglyphics evolved into an alphabet, syllables, and then into words. Words may involve the same neurons as those that store images. Children quickly learn that people and things have names. They identify a room, and learn the names of discrete objects in the room, such as books, ashtrays, tables, dishes, and pencils. They learn that everything has a name, and that the external world is identified by words. From the time a child is born until the age of 4 or 5, he or she looks at the external world and translates objects into words at a rate that will never be equaled again in his or her lifetime.

Iconography is the drawing of animals or objects. Logography is the use of signs to represent words, the best example being the Chinese language, which uses >4,000 characters. A syllabary is a collection of characters, each of which is a syllable. With the invention of an alphabet, a set of characters is created to correspond to the sounds of the language.

Research is being directed to the question of how the brain stores and continually updates images, words, and numbers to make sense of the enormous amount of sensory input that one is constantly bombarded with. We hope to determine how information is recorded and coded in sensory neurons, and how they transfer information to other regions of the brain.

In the late eighteenth century, graphs began to be used to show quantity, time series, and multivariate displays. Graphs were invented to make possible reasoning about quantitative information and replaced conventional tables of numbers. Graphs are a way to represent numbers as images, and they are often the simplest and most effective way to look at numbers. Again, we would like to know how numbers and graphs, showing complex ideas, are represented with clarity, precision, and efficiency in the brain.

Today, we can obtain four-dimensional (4-D) PET images of the regions involved in short-term and long-term memory. We can image the patterns of connectivity. Memories of words and images are stored in different locations in the brain and integrated with perceptions of objects being seen at the time. Images and words may be recorded in the brain by similar processes, with words being attached to the images. Words may be filed in the same regions of the brain as their associated images.

Molecular processes in the brain may be the basis of memory. We do not know which molecules are involved, or whether they reside in neurons or in synapses. Which synapses are involved? Are there specific patterns of the involved synapses? How are molecules within synapses involved in storage of memories? PET and single photon emission computed tomography (SPECT) can help answer some of these questions.

In addition to PET and SPECT imaging, optical imaging with near infrared light offers important advantages, the chief advantage being sensitivity, which can be translated into exquisite spatial resolution, far greater than that of PET or SPECT. MRI is being used in a multi-institutional study in six American cities, sponsored by the National Institutes of Health (NIH; Bethesda, MD). To study how the human brain changes as children grow up, because conventional radiological procedures are too dangerous or too invasive.

In the NIH study, 385 boys and girls were screened to be sure that they were free of illness, genetic predispositions to disease, toxic environmental exposures, or chronic health problems. They were screened periodically using structural magnetic resonance imaging to monitor the white matter, and magnetic resonance spectroscopy to monitor regional chemistry. Although the studies are ongoing, the data seem to indicate that healthy boys and girls do equally well on cognitive tasks but that boys are better at analyzing shapes and patterns. Girls perform mental activities faster than boys, and they have greater motor dexterity. It will take years before the study is completed. FDG studies with PET were not included in the study.

10.1. Over a Century Ago, William James Wrote

… "Our inner faculties are adapted in advance (by evolution) to the features of the world in which we dwell, adapted, so as to secure our safety and prosperity in its midst … Not only are our capacities for forming new habits, for remembering sequences, and for abstracting general properties from things, and associating their usual consequences with them exactly the faculties needed for steering us in this world of mixed variety and uniformity, but also our emotions and instincts are adapted to the very special features of that world."

"… mental action is uniformly and absolutely a function of brain-action, varying as the latter varies, and being to the brain-action as effect to cause.

… "The uniform correlation of brain-states with mind-states is a law of nature."

We are constantly surrounded by a world that we perceive, think about, and describe to others. Thought without speech is inconceivable. Nouns and verbs are thoughts. Not only things, but also concepts, such as truth, justice, courage, and divinity, are thought-words, a concept proposed by Arendt (1971). With PET, in theory, both thoughts and words can be physically localized in the brain.

We will some day learn what goes on in neurons during seeing and speaking. Images and language involve separate domains in the brain. Pictograms and hieroglyphics evolved into an alphabet, syllables, and then into words. Words are themselves images, and when words in the brain, they may involve the same neurons that store images. We would like to know how both words and images are stored in the neurons of the brain.

When one looks at a PET or SPECT scan, the first question that should be asked is, "What radioactive tracer was injected?" This tells us what molecular process is being shown in the images. The images are *functional* images, i.e., 4-D images often displayed as 2-D images, reflecting both the temporal and spatial distributions of the radioactive tracer. A computer is needed to construct and display the images. In an ^{18}F-FDG image of the brain, each picture element (pixel) reflects the rate of accumulation of the FDG tracer in different regions of the brain.

Color is often used to represent different molecular processes or quantitative parameters that indicate the temporal changes of the tracer in different brain regions. For example, a change from one color to another may represent a 5% difference in the rate of a molecular process being examined. Different colors do not represent boundaries between on anatomical structure, but instead quantitative differences in regional molecular processes. Different colors can be used to represent whatever one wishes to represent, and the viewer must know the relationships of the colors to molecular processes.

Another tracer technique was introduced for carbon dating of archeological specimens. Accelerator mass spectometry (AMS) is based on measuring atomic weight ratios of atoms in extremely small samples.

In 1966, in an article in NUCLEONICS, with TerPogossian, I wrote, "The most important radioactive tracer in biological research is reactor produced carbon-14, but it has never been widely used in nuclear medicine. The introduction of Carbon-11, Nitrogen-13, Oxygen-15 and Fluorine-18 revolutionized the study of regional chemistry in the living human body.

10.2. Mental Illness

A 73-year-old extraordinarily successful New York financier spent his entire adult life in a battle with depression. During his career, he had made many hundreds of millions of dollars, yet he struggled constantly with depression, often being counseled by the world's top physicians, psychiatrists and psychologists, and taking every type of antidepressant. He stopped reading the newspaper, which had been a favorite pastime. Unable to focus his attention, he picked up the paper and put it down over and over. He would go hours without talking to people.

He stunned his brother when he said, "I'm going to end my life and you will be the executor of my estate." His brother didn't believe that he would carry out the threat. He told his wife "I am tired of fighting." He refused hospitalization, saying it felt like "incarceration" to him. One month later, immediately after leaving his psychiatrist's office, he slipped away from his wife and jumped to his death from the ninth floor of his Manhattan apartment building (2006).

Nearly 20 million Americans suffer from depression, according to the National Institute of Mental Health. The disease has been described by physicians, scientists, and writers for centuries. In the fourth century BC, Hippocrates believed that mental illness was the result of an imbalance of four elemental body fluids-blood, yellow bile, phlegm, and black bile- in a context of four elements: air, fire, earth, and water. He viewed mental illness the same way that he viewed other medical disorders.

Mental illness is still viewed by most physicians as a medical illness, evidenced by the fact that drugs are widely used to treat patients suffering from schizophrenia, depression, mania, and anxiety. These medications often affect the levels of the amine neurotransmitters, norepinephrine, dopamine, and serotonin.

Genetic factors predispose a person to develop depression, interacting with other biological, psychological, social, and cultural factors. Abnormalities of specific genes have been correlated with an increasing risk of bipolar (i.e. manic-depressive) disorder. A genetic predisposition to develop schizophrenia is suggested by the fact that an identical twin of a person with schizophrenia has a 50% chance of developing the disease, even when reared separately by adoptive parents. Huntington's disease is related to a mutation of a single dominant gene. Some patients with senile dementia of the Alzheimer type have specific genetic abnormalities.

Molecular imaging is joining organ pathology, histopathology, microbiology, and biochemistry in the study and care of patients with mental illness. We can now examine the molecular processes involved in energy production and communication among the organs, tissues, and cells of the brain. Molecular imaging can address basic questions asked in the practice of medicine: What is wrong? What is going to happen? What can be done about it? Is the treatment effective?

10.3. Neuroreceptors

In 1973, Solomon Snyder and Candace Pert proved the existence of opiate receptors in the brain. Snyder had been a postgraduate student of Nobel Prize winner Julius Axelrod, at the National Institute of Mental Health in Bethesda, MD. Axelrod's research was concerned with the chemical events that occurred in the synaptic cleft that connects all neurons.

At Hopkins, Snyder and colleagues used radiolabeled naloxone, a drug that inhibits the pain-killing effects of morphine, to delineate the location of opiate receptors in the rat brain. Michael Kuhar used autoradiography to define these regions. This pioneering work made it possible to try to address the following important questions:

1. Will the knowledge of brain chemistry help understand how memory, emotions, and other mental functions are encoded in molecules in the brain?
2. How do molecular processes affect neuronal activity in determining mental activity?
3. Will the new knowledge of brain chemistry make life safer, healthier and happier?
4. Can we develop a mathematical model expressing how brain chemistry affects mental activity?
5. Can a better understanding of the relationship between brain chemistry and behavior help solve societal and political problems?
6. Will genetics and molecular imaging lead to the development of new drugs that make us less fearful and aggressive?

Centuries ago, Alexander Hamilton wrote, "People assume the pretext of some public motive, and have not scrupled to sacrifice the national tranquility to personal advantage or personal gratification ... Are not popular assemblies frequently subject to the impulses of rage, resentment, jealousy, avarice, and of other irregularities and violent propensities?"

Albert Einstein wrote, "Man has within him a lust for hatred and destruction. In normal times, this passion exists in a latent state, it emerges only in unusual circumstances, but it is a comparatively easy task to call it into play and raise it to the power of a collective psychosis." Is there a measurable brain chemistry of wisdom and virtue? Is there a darkness at the core of human nature, and, if so, can it be characterized and corrected chemically? Humans are strong intellectually but weak emotionally. We have difficulty in relationships with others who are culturally different. Intellectual activities are often impaired by emotions that ruin relationships with other people.

William James said, "We need to know a little better what are the molecular changes in the brain on which thought depends" (James, 1890).

"We gain our knowledge of the external world through the sensory experience of the objects in the environment, i.e., mental activity begins with sensory experience. To hate, love, think, feel, see, we must first perceive. All of our ideas are derived from these sensory impressions. We cannot envision the taste of a pineapple without having actually tasted one" (Locke, 1677)

Ralph Waldo Emerson said, "Nature is an endless combination and repetition of a very few laws ... If any one will but take pains to observe the variety of actions to which he is equally inclined in certain moods of mind, and those to which he is averse, he will see how deep is the chain of affinity (Emerson, 1926).

James wrote, "Between a mental and physical event, there will be an immediate relation, the expression of which, if we had it, would be the elementary psycho-physical law. What sorts of goings on occur when thought corresponds to a change in the brain? The brain is nothing but our name for the way in which billions of molecules arranged in certain positions may affect our senses ... The only realities are the separate molecules, or at most the cells ... The molecular fact is the only genuine physical fact ... The molecular facts correspond not to total thought but to the elements of thought ... The relation of the "known" object or feature, such as color, and the "knower" is infinitely complicated, and a genial, whole-hearted popular science way of formulating them will not suffice ... Where everything is change and process, how can we talk about state? ... Something definite happens when to a certain brain state, a certain consciousness corresponds ... A glimpse into what it is would be the scientific achievement before which all past achievements would pale. At present (1892) psychology is in the condition of physics before Galileo and the laws of motion, and before Lavoisier and the notion that mass is preserved in all reactions."

Molecular imaging is the foundation of molecular medicine. It is hard to imagine how a physician could diagnose or treat a patient with the disease of hypertension without measuring the patient's blood pressure; or care for a patient with an infectious disease without taking bacterial cultures; or care for a diabetic patient without measuring blood sugar levels. In mental illnesses, unless the patient is found to have a brain tumor or a stroke, the diagnosis is made on the basis of talking to the patient, performing subjective psychological tests, or finding abnormalities by neurological examination.

Structural imaging techniques, such as MRI, do not reveal functional or biochemical abnormalities. Measuring chemical processes in the brain in patients with psychological problems is likely to become as common as blood tests are today, determining what is wrong with the patient, what is likely to happen, what can be done about it, and is the treatment working. Simpler imaging devices will be developed and used to measure the effects of drugs on the brain.

Half a century ago, Konrad Lorenz wrote, "If we are powerless against the pathological destruction of our social structure, and if, armed with atomic weapons, we cannot control our aggressive behavior, this deplorable situation is largely due to our arrogant refusal to regard our own behavior as subject to the laws of nature and accessible to causal analysis ... we are still drawn by the same instincts as our pre-human ancestors ... Man is the most ephemeral and rapidly evolving of all species, but not the final achievement of creation. The long-sought missing link between animals and really humane beings is ourselves" (On Aggression, Bantam Books, 1963). This is still the challenge facing humankind today.

In his book *The Advancing Front of Medicine* (1941), George W. Gray wrote of his great admiration of Claude Bernard, a French physician from the late nineteenth century, who many call the father of scientific medicine: "The driving interest of Claude Bernard's life was the search for an understanding, in terms of physics and chemistry, of those processes, by which we live, by which we become ill, by which we are healed, and by which we die." If we cannot someday treat mental illness with the same effectiveness as we treat infectious diseases, we are all doomed."

Immanuel Kant (1724Ð1804), a major philosopher of the Enlightenment, said, "We do not know how far our observation and analysis of natural causes will lead us … Human beings are subject to all the laws prevailing in all adapted instinctive behavior."

Kant also said, "thoughts have robbed human beings of the security provided by his well-adapted instincts … We need to maintain, by wise rational responsibility, our emotional elegancies to cultural values. This need is as great as, if not greater than the necessity to control our other instincts."

Molecular imaging of the chemical processes in the human brain can define the chemistry of excessively aggressive behavior and identify the brain chemistry related to the instincts of friendship, love, kindness, and charity.

From time to time, there are paradigm shifts in the diagnostic process. In the 1990s, genetics showed how differences in the genetic makeup could reveal the risk of many diseases. Knowledge of these differences can form the basis for a new approach to earlier detection of disease and preventive medicine. Molecular medicine can detect disease at an earlier stage than is possible today, including mental diseases. Prevention will become more important than treatment. Mental illness, as is the case with cancer, will be characterized as the result of nature and nurture.

The three major domains of modern biological sciences are genetics, molecular imaging, and pharmacology. All three domains help produce an almost limitless supply of new, tailor-made drugs. Genetics and molecular medicine revolutionized drug design and development, with many drugs designed to alter the effects of defective genes. Pharmaceutical companies use gene splicing to produce specifically designed drugs. An example is *Gleevec*, designed to treat leukemia and stomach cancer. The drug blocks the action of an abnormal protein that tells the bone marrow to make abnormal white blood cells.

More than 100 oncogenes and 15 tumor suppressor genes have been linked to cancer. The same approach can be used in the characterization and care of biochemical abnormalities in the mentally ill. The "top-down" approach is to examine the psychological phenotypes of an individual and search for abnormal genes or molecular phenotypic processes. The "bottom-up" approach is to identify mutated genes and then search for their abnormal expression.

Molecular imaging successes in oncology and cardiology can be matched in preventing and treating mental illnesses. An example is the finding that thyroid cancer cells express the sodium/iodide symporter, an enzyme that concentrates radioiodine. The cancerous cells lose this ability as they become more malignant. They begin to avidly accumulate the radioactive tracer ^{18}F-FDG.

The general public must be told about the principles, practices, and opportunities available from molecular medicine. Health care is undergoing great change, and it is moving toward a solid foundation, based on biochemical and physiological processes. There is a shift in emphasis from treatment of overt disease to the early detection and prevention of future serious disease. The key is to better understand the pathological and biochemical changes that take place in the natural progression of disease, so that treatment can be effective.

We can now define mental diseases on the basis of molecular abnormalities in the brain. We need to learn the relationships among these different biochemical processes. We need a new taxonomy of disease, one that defines normal and abnormal molecular processes. We will search for antecedent manifestations of disease, at times viewed as the cause of a disease. For example, the increased accumulation of radioactive iodine by the thyroid gland is not the cause of hyperthyroidism, but a manifestation. The disease is caused by an antibody that mimics thyroid-stimulating hormone.

Electrical and molecular activities in neurons are the language of the brain, which writes the stories of the mind. Regional biochemistry in different parts of the brain can be classified in a multileveled hierarchy of semiautonomous subwholes, branching progressively into more specific smaller subwholes, that Arthur Koestler has called *holons*. The millions of biochemical processes in neurons can be classified into holons that include bioenergetic transformations, intermolecular communication, and structural integrity.

Organizing a system of disease classification based on molecular processes is the central problem. Hierarchies of holons can be built from molecular processes that are related to specific physiologic and pathophysiologic processes. On a foundation of genetics, molecular imaging can examine regional processes within the living human body in health and disease.

Regulations governing the use of radioactive tracers in molecular imaging must be greatly simplified. Because the mass of radioactive tracers is so small, toxicity is not a problem. By definition, radioactive tracers have no biological effect. Some are normal body constituents, such as ^{11}C-choline.

The Food and Drug Administration in the United States and the European Union Clinical Trials Directive are now reconsidering present regulations. Early human studies in drug development have been simplified. Integrated protocols involving multiple investigators prevent unnecessary delays in starting clinical trials and shorten the time to carry them out.

The use of surrogate biomarkers of disease also greatly reduces cost. In the words of Elizabeth Allen, Director of Scientific Affairs of Quintiles Limited, "Authorities understand the exploratory nature of early human studies and will consider novel approaches to study design supported by good science." The goal is to make increasingly large numbers of radioactive tracers widely available for clinical research, drug design, development, and regulatory approval, with the potential of reducing cost, in addition to greatly improving health care.

Molecular imaging can be used in well-designed and well-controlled studies of volunteers to determine whether there are patterns of brain chemical processes related to violent or criminal behavior. Such information could possibly help provide more effective supervision and services after the release of people from prison. An analogy is the measurement of blood pressure to monitor the effectiveness of drug and other treatments of patients with hypertension.

Patricia Godley, a former drug addict, stood before a town hall meeting on drug abuse and said, "You can open up all the jails in the world that you choose to, but if you don't get to the core of the human being that you are incarcerating, nothing is ever going to change, nothing" (Quoted by Torricelli, 1999).

A pessimistic view is that of David Brooks of *The New York Times*, who recently wrote, "As for today's white lab coat types who are trying to reshape the brain through drugs and neuroscience, Lemov has me looking at them in a new and much more skeptical way" (Lemov, 2006).

There is ample evidence that societal factors are involved in violent criminal behavior. For example, there is a close relationship between crime and the number of young men aged 16–24 in a society. In 1965, Daniel Patrick Moynihan wrote in *The Case for a Family Policy*, "A community that allows a large number of young men to grow up in broken families, dominated by women, never acquiring any rational set of expectations about the future—that community asks for and gets chaos. Crime, violence, unrest, disorder— most particularly the furious, unrestrained lashing out at the whole social structure— that is not only to be expected; it is very near to inevitable."

Drugs, prescribed or illicit, are also a cause of criminal activity. In 2005, 3.6 billion prescription drugs were purchased in the United States. Many of these drugs offer patients the possibility of modern day miracles. In 2004, 82% of the people in the United States reported taking at least one prescription drug, over-the-counter medicines, or dietary supplements in the week before surgery.

Thirty percent reported using five or more drugs, according to a study by the Sloan Epidemiology Center at Boston University. In 2005, among Americans over the age of 65, 75% took four prescription drugs every day. The same year, 1.6 million American teenagers and children took at least two psychoactive drugs.

Many persons taking prescription drugs believe that anxiety is as treatable with drugs as cough or insomnia. Excessive shyness can impair day-to-day functioning. Like depression and anxiety, it, too, is often treated with drugs, often unnecessary. Biological factors may lead to alcoholism, including genetic factors.

Neurotransmitter systems may be abnormal in persons suffering from alcoholism. The ability to image and quantify the regional concentrations and states of occupancy of neuroreceptors by drugs has had a major impact on the development of drugs used in the treatment of many types of mental illness. For example, the dopamine receptor has been the basic tool for screening potential drugs that affect the dopaminergic system. Almost every month a new peptide neurotransmitter is discovered. Many are localized in specific parts of the brain. The goal is to learn which mental functions are related to each neurotransmitter.

The greatest challenge facing the world today is the proliferation of nuclear weapons. There is no reason to believe that human beings will not use them under sufficient provocation and fear. Knowledge and materials for the building of atomic bombs are in the hands of criminals, as well as foreign countries and dangerous rogue states. William Langewiesche wrote (*Atlantic Monthly*, Jan/Feb 2006), "In July of 2000 Pakistan's Ministry of Commerce ran a full page notice in the English-language Pakistani press that advertised the nuclear weapons products that Pakistan had to offer: a full line of material and devices that ended just one step short of a ready-made bomb." The greatest threat to the human race comes from nuclear weapons. Dangerous materials to make nuclear weapons are sold by desperados all over the world, by using material obtained in Russia or Pakistan, or other countries.

The greatest fear is that rogue states will get their hands on nuclear weapons. Iraq is no longer a threat to develop nuclear weapons, but the present Iranian government intends somehow to produce or acquire weapons grade uranium, put together a quick atomic bomb, and threaten the United States or Israel. Despite their great efforts to prevent it, it may be impossible for Russia to control every gram of plutonium or weapons grade uranium. In 2002, North Korea received centrifuges for enriching uranium from Pakistan in return for missile technology.

A proposal has been made to have Russia to supply Iran with enriched uranium. In the former Soviet Union, between 900 and 950 locations store nuclear materials, plutonium, and uranium. Some of this material could be exported to Iran or other countries, even though the Russian and American governments are cooperating to prevent this from happening.

One way to detect subversive building of nuclear weapons is to monitor electric consumption looking for evidence of the operation of high-speed centrifuges. Another way is to track radioactive dust, which would suggest that uranium is being purified. There are distinctive chemicals, sounds, electromagnetic waves, and isotopes that can be monitored. A problem is Tehran's recent decision to restart its enrichment of uranium, a major step toward the production of nuclear arms. Under development are remote-controlled aircraft that can carry sensors to detect evidence of prohibited activity such as the detection of the ingredients such as uranium hexafluoride gas, which is processed in centrifuges during the enrichment process.

11

Violence and Crime

A psychologically normal 55-year-old man had a large acoustic neuroma removed from his brain. Following operation, he became so violently angry and aggressive that he was unable to work, and, according to his wife, became impossible to live with. After six months of misery, he was begun on treatment with a selective serotonin reuptake inhibitor (SSRI) to increase the levels of the neurotransmitter, serotonin, in his brain. After two weeks of treatment, he was able to return to his normal preoperative mental state, and go back to work.

The increased levels of serotonin in his brain resulting from the SSRI medication counteracted the effects of increased levels of epinephrine and norepinephrine. Fear is produced by epinephrine; anger by norepinephrine. The latter can result in excessive competitiveness, and lead to hate and violence.

Even jogging or running can lead to an addiction to epinephrine and norepinephrine, especially running long distances. Physical stress causes epinephrine to be released from the adrenal glands. Great quantities are released during extreme sports, such as bungee jumping, boxing, parachute jumping, skiing, or car racing.

Violent criminal activity seems to be associated with high levels of epinephrine, and may have played a role in the behavior of people, such as Charles Manson. He took the name *Helter Smelter* from a Beatles song of that name in the early 1970s. Manson was obsessed with the song and used its title as a symbol of rebellion, evil, ghoulishness, bloody violence, race war, and homicidal psychosis.

Whether one judges Charles Manson to be insane depends on the definition. He understood the difference between right and wrong, and today would be diagnosed as having an *antisocial personality disorder,* and in the past would have been called a *sociopath* or *psychopath.*

Manson's evil behavior and violence were related to heroin abuse and LSD. His followers, who he called his *family,* were strongly influenced by these drugs. They were among more than 300,000 Americans who abused heroin to "ease their troubled minds." Twelve million are addicted.

Over a century before, Marie Curie had expressed her hope that scientific knowledge would someday result in a decrease in violence, war, and poverty. We do not need to know how the brain works to be able to learn how drugs affect brain chemistry.

Genetic factors are clearly involved. The MAOA gene, located on the X chromosome, is involved in the metabolism of dopamine, noradrenaline, and serotonin. Low MAO activity behavior was the first gene strongly linked to violence and antisocial behavior.

In 2002, Avshalom Caspi, and a colleagues at the Institute of Psychiatry in London found that maltreated boys producing a variant of the MAOA gene were more likely to develop antisocial problems than were maltreated boys who had a variant producing high MAOA activity. Debra Foley, Ph.D., of Virginia Commonwealth University and coworkers reproduced these finding in 2004 (Caspi, 2004).

Kent Nilsson and his group at Uppsala University in Sweden have come up with the same finding, but with a twist. When adverse psychosocial conditions were taken into consideration, the short MAOA gene variant no longer produced antisocial behavior. Thus, the short gene variant affects antisocial behavior only under adverse psychosocial conditions.

Violent offenders in prison have low cortical brain blood flow and hypometabolism in nondominant frontal and temporal lobes, when compared to control subjects. Others have abnormalities in prefrontal regions. Some criminals have metabolic diseases, infections, tumors, malnutrition, poisons, trauma, or take drugs. Many have psychiatric problems, are poor, come from broken homes, or have suffered social ostracism that leads to resentment and hostility.

The law assumes that everyone has a free will, and is responsible for his actions. Yet, lawyers regularly present evidence that biological factors led their clients to criminal actions. More and more they use PET scans of drug addicts as evidence in court. It is a rule-of-thumb that whenever a new physical or chemical measurement of the human brain is developed, defense attorneys will try to use the findings in their clients to deny that they have free will.

Whether some of today' illegal, abused drugs should be legalized is often discussed. Traditionally, Canadians have been for legalizing these drugs, but the drug trade in the province of British Coumbia (BC) now generates $7 billon per year. There are 129 drug-related gangs in Canada, which have been increasing in number for decades.

The Senate Committee on Illegal Drugs in Canada promoted the legalization of marijuana, recommending that it be made readily available to persons over 16 years of age. It calls for amnesty for those previously convicted of marijuana possession, approximately 600,000 Canadians.

The issue is not whether the use of the drugs, such as marijuana, should be permitted in Canada, but whether laws should control these substances. Advocates of decriminalization cite economic factors. Opponents cite crime rates, and studies that show that marijuana can have lasting effect on the brain.

Those who oppose the legalization of marijuana fear that its use would be a stepping-stone to heroin, cocaine, or other harder drugs. Many young people believe it is a "harmless" drug, but, after a while, they want a bigger "high", and turn to heroin, LSD, or cocaine. Traffic accidents result from driving while "stoned", which is a problem. Poor judgment can lead to rape or robbery, and cause brain damage, cancer, lung damage, depression, or death.

Studies on Marijuana abusers with ^{18}F- FDG PET during performance of a memory task, revealed impaired neuronal activation in the middle temporal gyrus, the parahippocampal gyrus, and in the laterality of hippocampal activation (Bradley Voytek and David Geffen, UCLA). Within the first six months of decriminalization in California, arrests for driving under the influence of drugs rose 46% for adults and 71.4% for juveniles. The use of marijuana doubled in Alaska and Oregon when it was decriminalized in those states.

Justice Department figures in the United States show that approximately one-third of inmates used drugs prior to committing their crimes. A 1990 study published in the Journal of Drug Issues found that the more intoxicating the drug, the more serious the unfoward incident.

Nuyiek and Wemen write: "The Drug Enforcement Administration estimates that drug decriminalization would cost society more than alcohol and tobacco combined, perhaps $140–210 billion a year in lost productivity and job-related accidents.

"Keeping drugs illegal maintains criminal sanctions that persuade most people their life is best lived without drugs. Legalization, on the other hand, removes the incentive to stay away from drugs and increases drug use."

Trying to decide the question of the decriminalization of drugs nearly always raises the story of the effects of prohibition of alcohol. In 1920, the Volstead Act was passed by Congress. The whole nation became enforced teetotalers, the goal being to end all evils associated with drinking. Prohibition stopped the legal production, importation and distribution of alcoholic beverages, and these activities were taken over by criminal gangs, which fought each other for market control. Nevertheless, many people were sympathetic to bootleggers, and respectable citizens were regular customers of illegal speakeasies.

In the United States in 1921, there were 95,933 illicit distilleries, stills, and still works that were seized by government agencies. (Jane Lang McGrew). In 1925, the total jumped to 172,537 and up to 282,122 in 1930. In these seizures, 34,175 persons were arrested in 1921; by 1925, the number had risen to 62,747 and to a high in 1928 of 75,307. Convictions for liquor offenses in federal courts rose from 35,000 in 1923 to 61,383 in 1932. An illicit traffic in alcoholic beverages developed, from manufacture to consumption. The number of speakeasies throughout the United States ranged from 200,000 to 500,000. Doctors earned $40 million in 1928 by writing prescriptions for whiskey.

In 1923, there were 134 seizures of ships engaged in smuggling liquor. In 1924, there were 236. The Department of Commerce estimated that, in 1924, liquor valued at approximately $40 million was entering the United States annually. Section 29 of the Volstead Act authorized the production of fermented fruit juices. Although the intention was to save the vinegar industry and the hard cider of America's farmers, it was a boon to home winemakers as well. The grape growers of California produced a type of grape jelly suggestively called "Vine-go" which, with the addition of water, could make a strong wine within two months. The grapes were accompanied by the written warning: "Do not carry out the following steps, because, if you do, you will make wine, which is illegal."

In 1928, a dry-wet confrontation emerged in the presidential election between Alfred E. Smith, a Catholic New Yorker, and Herbert Hoover. Hoover solemnly praised the "great social and economic experiment" and tightened his grip on the dry vote.

The Wickerson Commission Report in 1931 stated: "There have been more sustained pressures to enforce this law than on the whole has been true of any other federal statute, although this pressure in the last four or five years has met with increasing resistance as the sentiment against prohibition has developed... That a main source of difficulty is in the attitude of at least a very large number of respectable citizens in most of our large cities and in several states is made more clear when the enforcement of the national prohibition act is compared with the enforcement of the laws as to narcotics. There is an enormous margin of profit in breaking the latter. The means of detecting transportation are more easily evaded than in the case of liquor. Yet there are no difficulties in the case of narcotics beyond those involved in the nature of the traffic because the laws against them are supported everywhere by a general and determined public sentiment."

On March 23, 1933, President Franklin Roosevelt signed into law an amendment to the Volstead Act known as the Cullen-Harrison Act, allowing the manufacture and sale of "3.2 beer" (3.2% alcohol by weight, approximately 4% alcohol by volume) and light wines.

Congress officially adopted a repeal by passing the 21st Amendment to the Constitution on December 5, 1933. The National Prohibition Reform Act declared that prohibition was "wrong in principle" and "disastrous in consequences in the hypocrisy, the corruption, the tragic loss of life and the appalling increase of crime which has attended the abortive attempt to enforce it." Drinking began at an earlier age.

McGrew states: "There is no single compilation of Prohibition statistics which would enable us to determine the degree of success which Prohibition enjoyed during its lifetime… Most observers conclude that the undertaking failed.

In spite of many sincere and determined efforts, no country in Europe or the Americas has yet succeeded in eliminating the use of alcohol by society by legislative fiat.

The consumption of alcoholic beverages was 73,831,172 gallons, or 0.6 gallon per person in fiscal year 1930 as contrasted with 166,983,681 gallons or 1.7 gallons per person in 1914.

The data published by the Department of Commerce in the Statistical Abstract of the United States reflect a different picture. The average annual per capita consumption of hard liquor from 1910 to 1914, inclusive, was 1.46 proof gallons.

The per capita rate for the Prohibition years is computed to be 1.63 proof gallons. This is 11.64% higher than the Preprohibition rate. Based on these figures, one observer concluded: "And so the drinking which was, in theory, to have been decreased to the vanishing point by Prohibition has, in fact, increased" (Tillitt, 1932). Others disagree with the implications of these unverified statistics.

According to McGrew, "there was a notable decrease in alcoholic psychoses and in deaths due to alcoholism immediately preceding the enactment of Prohibition and a gradual increase in alcoholic psychosis and in deaths from alcoholism in the general population after 1920."

The Website "Potsdam.edu" concludes that "National Prohibition not only failed to prevent the consumption of alcohol, but led to the extensive production of dangerous unregulated and untaxed alcohol, the development of organized crime, increased violence, and massive political corruption. Some people today insist that Prohibition was a success!

Stanton Peele has written: "Although consumption of alcohol fell at the beginning of Prohibition, it subsequently increased. Alcohol became more dangerous to consume; crime increased and became "organized"; the court and prison systems were stretched to the breaking point; and corruption of public officials was rampant. No measurable gains were made in productivity or reduced absenteeism. Prohibition removed a significant source of tax revenue and greatly increased government spending. It led many drinkers to switch to opium, marijuana, patent medicines, cocaine, and other dangerous substances that they would have been unlikely to encounter in the absence of Prohibition." With some uncertainty, one can conclude that prohibition was a failure. There was a decrease in alcoholism.

Today, the question is whether some illicit drugs, such as marijuana, should be decriminalized? The Dutch have long favored "harm reduction" rather than law enforcement to reduce substance abuse. Many coffee shops openly sell marijuana. They have banned the sale and cultivation of mushrooms, and the large-scale production of marijuana. Rotterdam has passed a law recently that will shut down nearly half of its grow shops. By the late 1990s the Netherlands had become the world's largest producer of Ecstasy, but in 2002 the government launched a five-year campaign against the drug.

Teenagers who said they had tried illicit drugs has fallen sharply since 2000, according to surveys by researchers at the University of Michigan. The percentage of students in 8th, 10th or 12th grades who tried methamphetamine declined by more than half over the same period, while cocaine abuse declined by almost a quarter among 8th graders and 10th graders.

Teenage abuse of other narcotics, including prescription drugs, is growing. The National Drug Threat Assessment reports that many more Americans over 18 are trying everything from heroin to marijuana to methamphetamine.

Legalization of mind-altering drugs represents a shift in European attitudes and laws affecting substance abuse. In Sweden only 12% of the adult population report that they have never taken abused drugs. The number of serious drug users decreased by 7% in 2007. In Switzerland, with a population of 7.5 million, 500,000 are thought to smoke marijuana occasionally. In 2004 a bill was introduced that would have decriminalized marijuana use by adults, but the bill was rejected. In 2007, 1.5 million British citizens were using marijuana. The UN's Office on Drugs and Crime has said: "The harmful characteristics of cannabis are no longer different from those of other plant-based drugs, such as cocaine and heroin."

Some believe that "full legalization is important, so that there can be sensible education about the possible dangers" (Boyce, 2007). With decriminalization, virtually overnight, the price of formerly controlled substances would plummet. All street crime, money laundering, gang violence, and the corresponding corruption in law enforcement that involves drugs, would disappear. The power of organized crime and drug cartels would decline drastically, with beneficial ripple effects throughout our society. The greatest improvement will be seen in impoverished communities. Street dealers will be gone. The length and cost of maximum sentences would decrease. Families would not be broken up by incarceration of parents.

"Based upon experience with the prohibition of alcohol, with decriminalization, there would be a temporary rise in drug abuse, which would eventually decline and level off, partly because of more robust and better-funded prevention programs (from the billions of dollars saved from a drug enforcement program that's no longer needed), and also because studies indicate there's a percentage of "addictive personalities" who will seek out drugs whether they're legal or illegal. Most people will not abuse recreational drugs even if they're decriminalized."

"Pharmaceutical companies would make safer substances to wean abusers off most addictive and psychoactive substances... Prohibition forces the commerce underground and makes it invisible."

PET or SPECT studies are often introduced in legal proceedings. Some of these cases involve head injury; others involve crime. Pre-existing brain damage militates against the death penalty. Although there are important issues of informed consent, well-designed studies could be designed to explore the possible relationship between brain chemistry and criminal activity. Violent crime might someday be viewed the same way as infectious diseases, where offending chemicals are analogous to offending micro-organisms. To be sure, environmental, social and preventive factors play a major role in controlling infectious diseases, but so also do antibiotics help afflicted persons, where treatment is based on identifying the infecting organisms.

By analogy, there may be measurable brain chemical abnormalities associated with a person's acting violently. For example, reduced levels of serotonin in the brain have been found to be associated with aggressive and violent behavior.

In persons committing suicide, the serotonin system is hypofunctional. The amount of 5-hydroxyindoleacetic acid (5-HIAA), the principal metabolite of serotonin, is low in the cerebrospinal fluid (CSF) of individuals attempting suicide (Stockmeir et al., 1998). Serotonergic activity in the postmortem brain is also low in suicide victims. Serotonin receptors (5HT1A and 5-HT2A) are increased in the prefrontal cortex, as a result of low amounts of bound serotonin. The hypofunctional state of the serotonin system causes disinhibition of excitatory systems mediated by neurotransmitters, such as norepineph-rine, leading to increased irritability and aggressive behavior. Mutated genes encoding serotonin synthesis and function are found in some violent persons.

Many addicts who end up in prison are mentally ill. Imprisonment for crimes resulting from taking or selling illicit drugs is not a solution.

"The easiest thing we can do is put people in jail, but you cannot prison-build your way into reducing crime" (Hynes, 2006). The U.S. prison population began to level off recently, but over two million adults are still in correctional institutions, an all-time high. There was a 4% rise in the number of persons in prison in the United States between 2000 and 2007, less than the 77% increase from 1990 to 2000. In New York, there are 12 mental health courts, which prescribe treatment of criminals rather than sending them to prison. There are 120 such courts throughout the country.

Will we someday monitor brain chemistry the way we take a person's blood pressure or measure blood cholesterol levels? We have begun to address these questions by examining the effects of drugs on brain chemistry, because it is a lot easier to learn how drugs affect the brain than to learn how the brain works.

What can be learned by carefully conducted research within the criminal justice system? In Oklahoma, more than a third of the state's 24,000 inmates have mental ill-nesses. According to the National Institute of Mental Health, one in four Americans over the age of 17 suffers from a diagnosable mental illness. About 45% have two or more mental disorders.

Crime and substance abuse are still widespread today, although since the early 1990s, crime has decreased in the United States. According to the National Crime Victimization Survey, violent crime and property crime fell significantly between 1993 and 2005. The FBI has reported that the overall rate of crime in 2007 was less than in the mid-1990s. Teenage drug use has declined significantly The use of Ecstasy has dropped by over 50% since the 1990s, methamphetamine by almost as much, and steroids by over 20%.

We have not reached nirvana. The gains are not yet secure, and could easily be lost. The social pathologies still affect us, so we must increase our efforts, both socially and scientifically.

Molecular imaging may be able to find molecular characteristics of both altruistic behavior and substance abuse. Discoverable brain processes may make a person want to help others, just as definable brain molecular processes may make a person susceptible to substance abuse. What are the relative roles of the chemistry of cognition and emotions in addiction? What are the relative roles of "nature" and "nurture? Cocaine is one of the most addictive substances, and strongly linked to crime. It prevents the re-uptake of dopamine, serotonin, and norepinephrine into presynaptic transporter sites. Inhibition of this re-uptake process elevates the synaptic concentrations of all three neurotransmitters, enhancing their activity on the mesocorticolimbic neuronal networks (Roberts et al). Cocaine also affects other neurotransmitter systems (opioidergic, glutamatergic, and GABAergic). Blocking D1 dopamine receptors with antagonist drugs is an important treatment of cocaine addition.

Alcohol is also related to criminal activities. Chronic alcoholics have long been treated with alcohol-deterrent drugs such as disulfiram (Antabuse) and calcium carbimide (Temposil), which can produce abstinence rates of more than 50%. These drugs alter the way in which the body breaks down alcohol. A person taking antabuse who consumes alcohol will have a violent physical reaction to the alcohol – nausea, vomiting and rapid changes in blood pressure occur.

11.1. Abused Prescription Drugs

In 1954, the Ciba Pharmaceutical Company (later called Novartis) introduced a drug called Ritalin (methylphenidate) that was originally used to treat depression, chronic fatigue, and narcolepsy. Beginning in the 1960s, it was used to treat children with attention deficit hyperactivity disorder (ADHD), known at the time as hyperactivity or minimal brain dysfunction (MBD). Investigators Joanna Fowler, Nora Volkov and their colleagues of Brookhaven National Laboratory showed that methylphenidate is a dopamine reuptake inhibitor, which increases the concentration of synaptic dopamine in the brain by blocking the transporters that remove it from synapses. Positron emission tomography (PET) showed that administering therapeutic doses of methylphenidate to healthy adult men increased synaptic dopamine levels.

During the mid seventies to early eighties, research showed that stimulant drugs improve the performance of most people, regardless of whether they have a diagnosis of ADHD, on tasks requiring good attention…this explains the high level of "self-medicating" around the world in the form of stimulants like caffeine and nicotine.

Drugs, such as cocaine, affect dopamine-producing neurons, blocking the dopamine transporter re-uptake sites that return synaptic dopamine to presynaptic neurons. In order for cocaine to make a person feel high, more than 60% of the brain's dopamine transporters must be blocked.

Both dopamine and norepinephrine re-uptake transporters bind methylphenidate, which has actions similar to amphetamine, calming patients with ADHD, reducing impulsive behavior and improving concentration. Adults with ADHD believe that methylphenidate increases their ability to focus on tasks and organize their lives. When doses of Ritalin comparable to those given to children to treat ADHD were given to normal adults, the oral doses did not result in a "high," in marked contrast to the effects of intravenous Ritalin, for which peaks are reached within eight minutes, and for cocaine, whose effects peak within five minutes after administration. Dopamine inhibits neuronal activity, which accounts for its "calming" effect on children with ADHD.

In the United States, methylphenidate is classified as a controlled substance, with medical value but also a high potential for abuse. People abuse MPH by crushing the tablets and snorting them to produce the "high." Ritalin has the same problem as cocaine or amphetamine in leading to possible addiction. As the number of children taking Ritalin has increased, it is at times over-prescribed to sedate "problem" schoolchildren to stop their disrupting class.

11.2. War

For over three thousand years, there has been almost continual warfare throughout the world, including the Greek-Persian war in the fifth century BC, the Peloponnesian War, Rome's battles in Carthage three centuries later, the Christian–Islamic wars of the Middle Ages, the Hundred Years' War between England and France, the American Revolution, World Wars I and II, the Korean War, the Vietnam War and now the Iraq war.

The world has been shaped by conquest of one nation by another. The *have nots* have continually preyed upon the *haves*. The invention of agriculture and the domestication of animals in the Stone Age led to the formation of tribes tribes. The conquest of Cuba by Diego Velazquez, Hernando Cortez and his hundreds of followers wiped out the entire empire of the Aztecs at the start of the sixteenth century. In 1526 and 1527, Francisco Pizarro and 160 other men brutally conquered the Incan empire in Peru, and stole immense hoards of gold and silver.

Destructive deaths have long been the curse of the human race. More than 50 million people have been killed over the past century. From 1915 to 1923 Ottoman Turks slaughtered up to 1.5 million Armenians. In mid-century, the Nazis liquidated six million Jews, three million Soviet POW's, two million Poles and 40,000 "undesirables." Four percent of the world's population was killed during World War II. US News and World Report (Jan/Feb, 2006) summarized the number of casualties in American wars: War of 1812 – 6,765 deaths; Mexican War – 17,435 deaths; Civil War – Union 646,392 deaths, Confederate 335,524 deaths; Spanish-American War – 4,108 deaths; World War I – 320,518 deaths; World War II – 1,077,245 deaths; Korean War – 139,858 deaths; and Vietnam War – 211,512 deaths. The total number of deaths was 2,759,357.

Tens of thousands of atomic and hydrogen bombs were built during and immediately after World War II, and 6,000 are now in existence, each having 20 times the destructive force of the atomic bombs at Hiroshima and Nagasaki. South Africa removed its nuclear weapons as part of the Nuclear Nonproliferation Pact (NPT). Today, forty nations throughout the world have the ability to produce nuclear weapons. The world barely escaped catastrophe in October 1962, during the Cuban missile crisis. There were 1,700 nuclear weapons in Cuba at that time. If the U.S. had invaded Cuba, it is too horrible to contemplate what might have been the consequences.

Mao Zedong killed 30 million Chinese, and the Soviet government murdered 20 million of its own people. In the 1970s, the Communist Khmer Rouge killed 1.7 million of their fellow Cambodians. In the 1980s and early 1990s, Saddam Hussein's Baath Party killed 100,000 Kurds. Rwanda's Hutu-led military wiped out 800,000 members of the Tutsi minority in the 1990s.

11.3. The Violent Brain

Molecular processes in the brain are related to violent behavior. These can be measured in the same way that we take a person's, blood pressure to detect hypertension, or measure blood sugar to detect diabetes. Excessive aggressiveness may someday be treated the way infectious diseases, diabetes, cancer or heart disease are treated. For years, sex offenders have been treated with drugs that increase serotonin levels in the brain, or with drugs that decrease testosterone levels.

At least 95% of all state prisoners will eventually be released. Many ex-inmates fail to make it on the outside, and return to prison. A current research project in New Jersey will provide intensive assistance to a group of ex-inmates and parolees, including anger management programs. Such programs should include molecular imaging studies to give them a better chance to succeed. The costs of the brain imaging research would be more than equaled by the potential costs of recidivism.

Michigan has increased funding for its Prisoner Re-entry Initiative to $33 million. In the program's two year existence, 14% of 8,000 released inmates who have received help have gone back to prison, a far lower number that the 48% of former inmates who returned to state prisons over a comparable period of time (Jones, 2007)

In 1962, Surgeon General Antonio Novello wrote: "If we are to succeed in stemming the epidemic of violence, we must first address the social, economic and behavioral causes of violence. We must try to improve living conditions for millions of Americans. We must try to provide the economic, and educational opportunities for our youths that racism and poverty destroy."

Expressing a different opinion, Surgeon General C. Everett Koop wrote: "Regarding violence in our society as purely a sociologic matter, or one of law enforcement, has led to unmitigated failure." What is needed is "major research on the causes, prevention and cures of violence, which might lead to new medical/public health interventions." It is time to examine these ideas by scientific studies.

Such studies would be controversial. Biological studies, particularly genetic studies, have in the past been controversial, opposed by some as efforts to make excuses for violence. Advocates of research on criminality believe that new scientific technology can help shed light on the biological correlates of violence and perhaps offer new solutions.

Previous investigators of the relationship between biology and crime have become involved in bitter controversies. The research as been called scientifically unsound and politically motivated. Opponents believe that it would lead to certain groups of people being branded as being prone to violence.

Neurotransmitters and hormones are involved in aggressive behavior. When men are stressed, aggression may be cathartic, while women may try to sooth the offender. This difference may be related to hormonal differences (Berman et al., 1993). Testosterone blood levels are ten times higher in men than women. These levels increase with success in a competitive athletic event or sexual intercourse. They decrease with failure.

Violence is the result of interactions of innate genetic factors and environmental factors. For example, criminals in Taipei had abnormalities of the gene encoding the 5-HTT (serotonin transporter) related to violent criminal behavior (Liaoa, 2004). Serotonin (5-HT) and dopamine (DA) transporters in violent criminals were found to be less than in sex-matched healthy controls, when studied with single-photon emission tomography and iodine-123-carbomethoxy-4-iodophenyl tropane (^{123}I-CIT) (Tiihonen et al., University of Kuopio, Finland). Richard J. Davidson et al. of the University of Wisconsin showed that the serotonergic projections in the prefrontal cortex are dysfunctional in persons who are impulsively violent.

The predominant effects of serotonin are inhibitory. It enhances the action of the GABA, the major inhibitory neurotransmitter. Serotonin has an excitatory effect by increasing the release of dopamine, a neurotransmitter involved in love, sex, and feelings of well being. Reduction in serotonin activity makes people incapable of experiencing normal positive rewards. Dopamine activity is enhanced by alcohol, cocaine, amphetamine, heroin, and risk-taking behavior.

Bernhardt (1997) suggested that elevated testosterone alone doesn't account for aggressive behavior. He believes that testosterone is linked more to dominance in general than to aggression. Bernhardt speculates that low serotonin activity is associated with increased responsiveness to aversive stimuli. High testosterone levels encourage dominance-seeking behaviors, "which put the individual into situations in which frustration of dominance can occur." When this happens, low serotonin levels result in a greater likelihood of an intensely negative emotional reaction, that is, a greater chance of aggressive behavior.

The hypothalamus and amygdala, "prominently associated with both testosterone and serotonin," play a key role in aggressive responses to situations in which efforts at dominance are frustrated. He notes that "low serotonin levels have been found in the hypothalamus and the amygdala in aggressive animals," and that "testosterone action in both of these brain structures has been shown to increase aggression in various animal species."

Women prisoners who had committed violent crimes had higher levels of testosterone than those prisoners who committed nonviolent crimes (Dabbs, Ruback, Frady, Hopper and Sgoritas, 1988). Increased testosterone levels may be the result of aggression rather than its cause (Zuckerman, 1991).

Boys have more trouble than girls with teachers, parents, and classmates. Later, they are more likely to abuse alcohol and drugs (Dabbs and Morris, 1990). Aggression is often related to alcoholism. Serotonin, dopamine, epinephrine and cortisol are also involved in aggression, impulsivity, and suicidal behavior.

Education, social status and other environmental factors, past and present, are reflected in biological characteristics. Leonote Tiefer, a physiological psychologist, has written: "Social experience affects brain chemistry."

Fever and weight loss in a young man may be related to his living in slum housing, having a poor diet, and being exposed to other sick people, but his diagnosis and treatment will be based on the finding of tubercle bacilli in his sputum. He reacts to tubercle bacilli in a way different from others. Similarly, in different people living in different societies, brain chemical processes may be different. The brain does not respond in an all-or-none, on-or- off, yes-no fashion. There are demonstrable differences in various parts of the brain.

Hostility, aggressiveness and violent behavior involve the hypothalamus, limbic system, and brain stem. At Harvard, Walter Cannon showed that removal of the cerebral cortex in experimental animals led to anger when the animal was slightly provoked, which he called "sham rage." Phillip Bard in the Department of Physiology at Johns Hopkins carried out intensive studies of this phenomenon while I was a medical student.

11.4. Food for Thought

Glucose is the principal source of energy for the neuronal activity in the brain. ^{18}F-FDG studies of the human brain make it possible to identify regions of the living human brain when neuronal activity is increased. Positron emission tomography (PET) studies with ^{18}F-FDG were performed in 41 murderers, who had pleaded not guilty by reason of insanity. They were found to have reduced glucose metabolism in their prefrontal cortex, superior parietal gyrus, left angular gyrus, and corpus callosum. Asymmetry was found in the amygdala, thalamus, and medial temporal lobe (Adrian Raine, Monte Buchsbaum, and Lori LaCasse Biological Psychiatry, Vol. 42, 1997).

Adriane Raine said that "...this is the largest sample of violent offenders assessed by functional brain imaging...Damage to the prefrontal cortex can result in impulsivity, loss of self-control, immaturity, altered emotionality, and the inability to modify behavior, which can all in turn facilitate aggressive acts." Although prefrontal dysfunction may be common to many mental disorders, abnormalities in other brain areas help define other neuronal networks leading toward violence ("Brain abnormalities in murderers indicated by positron emission tomography," Adrian Raine, Monte Buchsbaum, and Lori LaCasse, Biological Psychiatry, Vol. 42, 1997).

Using FDG-PET, others have found that murderers had poor functioning of the prefrontal cortex part of the brain – the area believed to control and regulate aggressive behavior (Brain abnormalities in murderers indicated by positron emission tomography. Raine A, Buchsbaum M, LaCasse L. Biol Psychiatry. September 1997,15;42(6):495–508)

PET studies with FDG have shown that neuronal activity is increased in the amygdala when a person anticipates an electrical shock. Activity increased in the right amygdala of subjects viewing fear-inducing human faces. Morris and colleagues (PNAS 96, February 16, 1999) suggest that a subcortical pathway to the right amygdala, via the midbrain and thalamus, affects vision.

Hostility and violent behavior increase glucose utilization in the limbic system, temporal cortex, amygdala, hippocampus, and hypothalamus. The orbital frontal cortex and anterior cingulate cortex are involved in the inhibition of aggressive behavior.

Discovering biomarkers of a tendency to violence and crime is now possible. As William James stated, we need to direct our energies to create "the moral equivalent of war." He meant the war against violence. Chemical *biomarkers* measured by molecular imaging can result in better for diagnosis, treatment, and prevention.

Pain, anger and aggression are often linked. Anger is a feeling; aggression is its expression. Both are instinctive responses to threats and frustration of one's goals and desires. One can act aggressively without feeling anger, for example, soldiers in battle. The hypothalamus and limbic system, primitive brain structures, are involved in emotional behavior, including rage and fear. The limbic system, located on the medial side of the temporal lobe, includes the olfactory cortex, amygdala, and hippocampus. The amygdala is involved in fear and the fight or flight response.

A sudden loud noise can cause the adrenal glands to release adrenaline, which signals the heart, blood vessels and liver to carry out innumerable chemical processes, including the release of glucose from the liver, and insulin from the pancreas. When the perceived danger has passed, different molecules are released to turn off the excited adrenal cells.

Patients with major depression often have increased neuronal activity in the amygdala, which in turn causes increased neuronal activity in the autonomic nervous system. The amygdala fails to inhibit the hypothalamus. "People who show greater activation of the amygdala and cingulate cortex during a demanding cognitive task show a greater rise in blood pressure" (Gianaros, Peter et al, NeuroImage, 35, 795–802, 2007).

In addition to involvement of the hypothalamus and autonomic centers in the brain stem, including the vagal nuclei and sympathetic nervous system, the amygdala is also connected to the frontal cortex, mediodorsal thalamus, and medial striatum.

Activation of the anterior hypothalamus is associated with a person's being calm and peaceful. The posterior hypothalamus is associated with rage and aggressiveness. The brain stem controls autonomic functions that regulate the heart, lungs, and intestinal tract, processes that are involved in the response to stress.

11.5. Lead Toxicity

Rick Nevin believes that lead poisoning accounts for much of the violent crime in the United States (Washington Post, July 8, 2007). Exposure to lead as a child correlates with violent behavior later in life, according to studies of the association between lead poisoning and crime rates in nine different countries. "Sixty-five to ninety percent or more of the substantial variation in violent crime in all these countries is explained by lead." Lead in U.S. paint and gasoline fumes has exposed toddlers to lead toxicity because they put their contaminated hands in their mouths. He believes that lead toxicity is not the only factor leading to crime, but is a big factor.

Deborah Denno is one of the most influential women lawyers in America, and has taught at Fordham Law School since 1991. She also argues for a link between lead and crime (San Francisco Chronicle, September 26, 2007). She explored 3,000 factors in 1,000 children who were followed from birth to age 22, and found a highly significant link between observed lead toxicity and the likelihood of criminal activity. Lead poisoning was the best predictor at age 7 for aggressive behavior in school, juvenile delinquency, and eventually criminal violence.

In identical twins, genetic and environmental factors each account for 50% of the differences in memory (Dominique de Quervain. Nature Neuroscience, 2007). Monozygotic twins raised apart from an early age were examined for signs of antisocial personality disorder. In male twins, genes accounted for 47% of direct physical aggressiveness, 40% for indirect physical aggressiveness, and 28% for verbal aggression. Direct physical aggression is violence toward others; indirect aggression is aggression toward objects; verbal aggression is aggression expressed in speech.

11.6. Violence and Serotonin

In 1996, Lesch and by Ogilvie reported that mutation of the gene (SLC6A4) that encodes the human serotonin transporter, makes a person anxious and afraid (Lancet 347, 731, 1996). Each parent passes to his or her offspring either a short or a long version of this gene. The short version encodes less efficient transporters. People who inherit one (or two) of these short forms suffer from anxiety. The aggressiveness of some patients with Alzheimer's disease may also be related to these mutations.

In 1976, Asperg and others found reduced levels of the serotonin metabolite 5-HIAA in the cerebrospinal fluid of depressed patients who had made suicide attempts (Arch Gen Psychiatry 1976;33:1193–7). Brown and others also found a relationship between aggression, history of suicidal behavior, and 5-HIAA in cerebrospinal fluid (Psychiatry Res 1979; 1:131–9. 29).

5-HIAA in the cerebrospinal fluid (CSF) was low in impulsive, violent offenders. (Linnoila, Life Sci 1983;33:2609–14.). Low levels of CSF 5-HIAA were also found in impulsive arsonists and other violent criminal offenders.

Fenfluramine was introduced in the U.S. in 1973. It increase serotonin. The drug was withdrawn from the U.S. market in 1997 after reports of heart disease and pulmonary hypertension. Serotonin may be involved in depression by regulating the proliferation of cells in the hippocampus. Serotonin agonists (fenfluoramine) promote cell proliferation.

When aggressive patients were given fenfluramine, there was increased metabolic activity in parieto-temporal cortical regions, and decreased uptake of the tracer in the anterior cingulate. Impulsivity was correlated with increased regional glucose utilization in the superior and middle frontal cortex.

Positron emission tomography was used to measure regional glucose utilization in the brains after administration of meta-chlorophenylpiperazine (m-CPP), a serotonergic agonist that stimulates serotonin receptors (Antonia New et al, Arch Gen Psychiatry. 2002;59:621–629).

Unlike normal subjects, patients with impulsive aggression behavior did not activate the left anteromedial orbital cortex when given this drug. The anterior cingulate was deactivated when the drug was given to aggressive persons, while the posterior cingulate gyrus was activated in aggressive persons. The latter became inactive in control subjects. Persons suffering from impulsive aggressive behavior did not have the normal increase in glucose utilization in the left antero-medial orbital frontal cortex. Thus, impulsive aggressive behavior seems to be related to decreased serotonin effects in certain brain regions.

Borderline personality disorder (BPD) and major depressive disorder (MDD) are associated with low serotonergic activity (Coccaro et al, 1989; Oxenkrug, 1979; Siever et al, 1984). Serotonin abnormalities are also found in persons attempting suicide (Asberg, 1997), as well as in persons with impulsive aggression (Brown et al, 1979).

Serotonin reuptake inhibitor drugs (SSRIs) can make some people violent. Glaxo Smith Kline, the manufacturer of the SSRI, paroxetine, reported 44 aggressive events in 11,491 patients taking the drug. They warned that clinicians need to be aware of these possible effects, but stated that serious violence is rare. Some people with increased serotonin levels in the brain for a long period of time may develop problems in modulating anger, depression or impulsive behavior.

11.7. Dopamine

The mesolimbic dopaminergic system is the principal "reward" pathway in the brain. Amphetamine, cocaine, and opiates cause dopamine release in the nucleus accumbens. Food seeking and sexual behavior do so as well.

Dopamine is synthesized from its precursor, L-dopa in the ventral tegmental area (VTA) and in neurons that project to other regions, including the nucleus accumbens. From the ventral tegmental area (VTA axons from dopaminergic neurons project into the posterior dorsal caudate nucleus and putamen. Dopamine is involved in mental activities, including motivation and reward-seeking, as well as motor activity.

Dopaminergic axons project into the prefrontal cortex and amygdala, a region of the brain associated with fear. Drugs of abuse, including nicotine, marijuana, alcohol, cocaine and heroin, bring about increased secretion of dopamine in the brain. These chemical effects can last for months or even years after a person stops taking the abused drug. Craving for the abused drug often leads to compulsive drug seeking.

11.8. Acetylcholine

In 1926, Otto Loewi discovered that acetylcholine was the principal neurotransmitter of the parasympathetic nervous system. As with other neurotransmitters, acetylcholine is synthesized and stored in vesicles within presynaptic neurons. Ninety percent of the cholinergic neurons in the brain are muscarinic, controlling salivation, sweating, dyspnea, diarrhea, vertigo, confusion, weakness and coma. Acetylcholine binds to muscarinic receptors on smooth muscle cells, innervated by post-synaptic fibers of the parasympathetic nervous system. The thalamus and cerebellar cortex have nicotinic receptors, while most other regions of the brain have muscarinic neurons.

In 1936, Loewi received the Nobel Prize with Henry Dale. They discovered that acetylcholine has two types of action. The drug, atropine, produces the "muscarinic" effect of acetylcholine by binding to muscarinic acetylcholine receptors. Acetylcholine also has "nicotinic" actions, stimulating the autonomic nervous system and voluntary muscle fibers. Nicotine stimulates the noradrenergic system. It acts by opening ion channels across the membranes of post-synaptic neurons, which generates electrical action potentials.

Jean Nicot, the French ambassador to Portugal who took tobacco leaves imported from America to Catherine de Medici to be used to treat migraine, coined the word *nicotine*. It affects blood vessels, sweat glands, piloerectal muscles of hair, the eye, heart, intestinal tract, spleen, bladder, lungs and sex organs. The autonomic nervous system constitutes a small fraction of the total number of neurons in the human brain, accounting for no more than a million neurons, compared to more than 10 billion neurons in the neocortex alone (McGeer et al. 1978).

Nicotine patches exemplify efforts to modify mental activity. Patches are also being developed to treat cocaine abuse, in which the brain is flooded with dopamine and other neurotransmitters. Craving occurs when the brain reacts by diminishing the effects of cocaine. More and more cocaine must be taken to overcome the tolerance and re-create the high.

Soloff and others found that serotonin in the prefrontal cortex may be involved in impulsive aggression. Studies with PET of aggressive behavior in subjects without psychiatric or criminal histories implicate the orbitofrontal cortex, the right anterior cingulated cortex, and the bilateral temporal lobes.

11.9. Glutamate

The neurotransmitter, glutamate, is a major excitatory neurotransmitter. Glutamate receptors are called ionotropic receptors. The effect of glutamate is to cause charged ions such as Na^+ and Ca^{2+} to pass through a channel in the centre of the receptor complex, which results in a depolarization of the plasma membrane and the generation of an electrical current that is propagated down the dendrites and axons of the neuron. Kandel found that cyclic AMP increased when these effects were retained in the neurons, to provide a chemical foundation of short-term memory. Cyclic AMP activates an enzyme, protein kinase, which acts like a neuronal switch. Serotonin increases this reaction.

Kandel then proceeded to try to use this new information to develop memory-enhancing drugs. In 1996 he and his colleagues founded *Memory Pharmaceuticals*, a publicly traded company to develop drugs based on Kandel's AMP work. Roche joined in this work but discontinued their collaboration after two years. The problem was that the new drugs being developed had a devastating effect on the prefrontal cortex, which is involved in attention, working memory for things such as telephone numbers and addresses, and executive functions such as organizing and planning. A drug that may affect a part of the brain, such as the hippocampus, may also have a global effect, in this case, on the prefrontal cortex, dentate gyrus, and amygdala. Memory does not reside in a single region of the brain. The hippocampus is the site of storage of long-term memories of experiences. The so-called "semantic" memory for facts and figures relies less on the hippocampus. With age, one's "declarative" memory – where you saw someone, what someone told you, and other facts about your life – become less efficient.

11.10. Effect of Diet

High fat diets are believed to have an effect on mental activity. Omega-3 and omega-6 fatty acids can affect mental activity and behavior, not only in normal people but also in patients with mental illness. Fatty acids not only serve as an energy source, but also play a role in the formation of neuronal membranes.

Neuronal membranes contain phospholipids. A lack of essential fatty acids, such as omega-3 fats, or excessive intake of saturated fats, margarine, cholesterol, and animal fatty acids, can result in abnormalities. The neuronal cell membrane regulates the passage of molecules into and out of the cell. Neuronal membrane fluidity is believed to impact behavior, mood, and mental function. Physical properties, including the fluidity, of neuronal membranes affect neurotransmitter synthesis, signal transmission, uptake of serotonin and other neurotransmitters, neurotransmitter binding, and the activity of key enzymes that break down neurotransmitters like serotonin, epinephrine, dopamine, and norepinephrine.

EPA (eicosapentaenoic acid) and DHA (docosahexaenoic acid) are polyunsaturated fatty acids that are part of the omega-3 family. Essential fatty acids include Omega-3 and Omega-6 fatty acids. Omega-2 and Omega-6 are involved in the synthesis of pros-taglandins.

Barry Sears has promoted the idea that the increasing incidence of depression in modern life is related to a low intake of fish and fish oil. Depression leads to an increasing intake of serotonin-boosting drugs, including Prozac, Paxil, and Zoloft. He postulates that many people of the United States have a serotonin deficiency. SPECT studies of regional cerebral blood flow show that high-doses of dietary fish oil improve cerebral blood flow.

Depression in Japan is far less common than in America. New Zealanders have 50 times the rate of depression as the Japanese, and eat less fish than anywhere in the industrialized world. They eat very large amounts of harmful omega-6 fatty acids. In Greenland, Eskimos, who consume 7–10 g of long-chain omega-3 fatty acids every day, are rarely depressed, even though their living conditions can be pretty depressing because there is only an hour or two of sunlight during the winter months. Perhaps the effects are not primarily due to diet, but are related to the Japanese and Eskimos not being genetically predisposed to depression. The amount of fish they consume may have nothing to do with the incidence of depression. The ratio of two essential fatty acids, arachadonic acid and eicosapetenoic acid) is highly elevated in the cerebrospinal fluid of depressed patients when compared to nondepressed patients."

Neuronal membranes contain large amounts essential fatty acids, derived entirely from the diet. Animal studies demonstrate a significant increase in the amount of serotonin in the frontal cortex of animals that consume a high fish diet compared to a standard diet rich in omega-6 fats. Andrew Stoll and his colleagues at Harvard Medical School reported that lithium and valproate, used to treat depression, block the release of arachidonic acid.

Joseph Hibbeln of the National Institutes of Health examined the levels of omega-3 fatty acids in the blood of 50 patients hospitalized after attempting suicide. Normal persons with high blood concentrations of eicosapentaenoic acid (EPA) had fewer psychological traits related to suicidal risk. He suggested "some subgroups of suicidal patients may reduce their suicidal risk with the consumption of EPA." Another study showed that dietary intake of EPA and DHA influence serotonin-related behavioral functions.

11.11. Terrorism

Former Senator Sam Nunn has said the detonation of a small, crude weapon that inflicted less damage than that produced at Hiroshima would result in "psychological damage that would be incalculable." In the 1990s, there were 20,000 nuclear warheads and stockpiles for uranium and plutonium capable of making another 40,000 warheads in the Soviet Union. Large quantities of uranium are stored in more than 130 civilian nuclear reactors around the world. A year after the collapse of the Soviet Union in 1990, the United States provided hundreds of millions of dollars annually to help dismantle Soviet nuclear weapons and improve security at the nuclear sites. Since 1991 a Cooperative Threat Reduction Program has spent $10 billion to inactivate these weapons. Half have been inactivated, but 134 tons of plutonium remain in storage sites.

Saudi Arabia, Egypt, Morocco and Algeria are contemplating developing nuclear power, and South Africa, Brazil, Canada, Argentina and Australia are considering uranium-enrichment programs. Mohamed ElBaredei, Director General of the International Atomic Energy Agency has said: "…if a country is capable of doing that, they are virtually a nuclear weapons state.

Konrad Lorenz believed that "man's aggressive behavior as manifested in war, crime, personal quarrels, and all kinds of destructive and sadistic behavior is due to a phylogenetically programmed, innate instinct which seeks for discharge and waits for the proper occasion to be expressed." He did not believe that violence stems from human's animal nature, but is shaped by culture and social forces. He deemphasized emotions and emphasized social factors as determinants of behavior. A person's passion to love or to hate depends on societal and cultural factors in a biological setting.

Current research tries to relate brain biochemistry to love, hate, ambition, greed, jealousy, envy, and aggressiveness, hoping to define the importance of instincts, historical, social, and cultural factors. Such knowledge can help eliminate harmful, life-threatening behavior. Fear is an ever-present emotion, especially when people face what they can't control. They do not want to be made unwitting targets of political risks.

In passing from the eighteenth to the nineteenth century, knowledge decreased fear. Intellect and judgment made it possible to control emotions. Democracy replaced aristocracy. Human beings rebelled against authority and became free. Human beings began to be viewed as good by nature. Suppressed instincts of altruism, innate consciousness of right and wrong, love and honor, pity and pride, compassion and sacrifice were passed on to future generations, because of their survival value. In the words of William Faulkner, winner of the Nobel Prize in 1950:

"I believe that man will not merely endure. He will prevail. He is immortal, not because he alone among creatures has an inexhaustible voice, but because he has a soul, a spirit capable of compassion and sacrifice and endurance."

Many questioned the theological foundations of morality. Voltaire and Rousseau believed in the moral necessity of religion, concluding that the loss of religion would be fatal to maintaining morality among the masses of humanity. Some, such as Marquis deSade, however, taught that all desires are good, and all moral distinctions were delusions. Morality would be impossible without supernatural rewards and punishments.

In 1905, Inazo Nitobe wrote: "beneath the instinct to fight, there lurks a diviner instinct – to love" (Bushido, 1969). In his 1988 book, The Inner Reaches of Outer Space, Joseph Campbell wrote: "Whether in the depths of the forgotten sea out of which life originated, or in the jungle of its evolution on land, or now in these great cities that are being built to be demolished in recurrent wars, the same dread triad of the god-given urgencies of feeding, procreating, and overcoming are the motivating powers."

Nobel Peace Prize winner Elie Wiesel said: "Only human beings can move me to despair, but only human beings can remove me from despair." It behooves neuroscientists to "try to get inside the heads of the monsters who did this." History can be shaped by individuals, and not just by forces beyond our control.

11.12. Islam

Islam advanced while the West declined during the Middle Ages. By 1928, Constantinople had a population of 2 million, which was twice that of London, three times that of Paris, and eight times that of Rome. The Sunni and Shi'a sects divided Islam in the same way that Catholics and Protestants divided Christianity. A third sect, Wahhabism, was founded by Mohammed ibn-Abad-el-Wahab in what is now Saudi Arabia. The Sunni were the orthodox; the Shi'a were the heretics; and the Wahhabites were the Puritans of Islam. Wahhabism remains the dominant form of Islam in Saudi Arabia, Kuwait, and Qatar, with each person interpreting the Koran for himself.

Wahhabism rejected the "moral decline and political weakness" in the Arabian penninsula. In the battle against the science and philosophy of the Christian Middle Ages, industry was thought to be inferior to art. The domes and tiled minarets were preferred over the spires and arches to Gothic cathedrals. In the eighteenth century, Iranian civilization was thought to be superior to contemporary Western cultures. A Moroccan traveler, returning from Europe, exclaimed "What a comfort to be getting back to civilization (Rousseau and Revolution, W. and A. Durant, 1967).

Osama bin Laden justifies violence on religious as well as political grounds. Muslims are obliged to call nonbelievers to Islam, and defend the Muslim community from attack. War against Christians and Jews is the duty of every Muslim. Osama bin Laden said years ago that he had the right to use nuclear, biological and chemical weapons in carrying out his jihad.

Garry Wills has addressed the question: What Is A Just War? (New York Review of Books, Nov 18, 2004). When is it permissible to kill other human beings? The first rule is that one must discriminate between combatants and noncombatants. Terrorism – the killing of innocent people to make a political statement or drive political actions – is never justified.

Clausewitz said that war is fueled by emotion, which always outruns its intent with a constant ratcheting up of hatred. Hate produces atrocities, which provokes answering atrocities in an upward spiral. Thucydides said: "War...is a tutor of violence, hardening men to match the conditions they face. Suspicion of prior atrocities drives men to surpass ...their own cruel innovations, either by subtlety of assault or extravagance of reprisal." To St. Augustine, even a just war is a fountain of evil: "Anyone who looks with anguish on evils so great, so repulsive, so savage, must acknowledge the tragedy of it all; and if anyone experiences them or even looks on at them without anguish, his condition is even more tragic, since he remains serene by losing his humanity."

Fear, anger, and aggressiveness are encoded in human genes, passed on from generation to generation of animals who needed to kill to survive. Altruistic genes promoting love and care of offspring continued to evolve as human laws and morality increased one's probability of reproducing offspring.

The advanced societies of the Islamic world preceded those of the Western world. Courtly love, algebra, paper, and the abacus were passed from the Middle East to Europe. The beautiful architecture of Islam, with its pointed arches, arrived via Spain, Salerno and Sicily. The Islamic World evolved the idea of the university as a place of learning in which students congregate. An example is the Al-Azhar University in Cairo. Throughout the Middle Ages, Christians and Muslims met together to trade and study as often as they met on the battlefield.

Some believe that following religious principles can prevent war. The Crusades are often used to show the evil of religion, but they were an extension of barbaric invasions, not primarily an effort to recapture Christian churches lost to Islam.

The Byzantines of Constantinople and the caliphate of Damascus carried abroad the best characteristics of Greek and Roman culture long after their civilizations had been destroyed in Europe. The learning, customs, architecture and literature of Islam were as great as that in the West in its best days.

Persons, such as Bernard Lewis (From Babel to Dragomans: Interpreting The Middle East, Oxford U. Press, 2007) portray a hostile Islam, hell bent on the conquest and conversion of the West. He wrote: "The struggle between these rival systems has now lasted for some fourteen centuries. It began with the advent of Islam, in the seventh century, and has continued virtually to the present day. It has consisted of a long series of attacks and counter attacks, jihads, and crusades, conquests and re-conquests." His ideas are the intellectual foundation of present-day neoconservative thinking. Islamic hostility to America is manifest by *envy* and *rage*, directed against an ancient rival, and derived from feelings of humiliation, a growing awareness of having been overtaken and overwhelmed by those whom they regard as their inferiors. For a thousand years, Islam was technologically superior to Christendom and dominated its Christian neighbors, but since the failure of the Ottomans in 1683, the Muslim world has been in retreat.

Lewis states: "This is no less than a clash of civilizations, the perhaps irrational but surely historic reaction of an ancient rival against our Judeo-Christian heritage, our secular present, and the worldwide expansion of both." In contrast, most Americans believe that the solution to the problem of terrorism is to prevent the jihadists from attracting wide support in the Muslim world, to call for the Muslim majority to act against the ideology of the terrorists. Middle East scholar Daniel Pipes says that "radical Islam is the problem, and moderate Islam is the solution" (Joshua Muravchik and Charles Szom, Commentary, Feb 2008).

Debates about the morality of war have been going on since the time of the Prophet Muhammad. President Bush and Prime Minister Blair have said that terrorists put themselves beyond the bounds of Islam by killing innocent people. After the attacks in New York on Sept 11, 2001, President Bush said: "These acts of violence against innocents violate the fundamental tenets of the Islamic faith…the face of terror is not the true faith of Islam. That's not what Islam is all about. Islam is peace." These sentiments were echoed by Prime Minister Tony Blair: "There is nothing in Islam which excuses such an all encompassing massacre of innocent people, nor is there anything in the teachings of Islam that allows the killing of civilians, of women and children, of those who are not engaged in war or fighting."

Osama bin Laden and his associates have cited the requirement that nonislamic enemies of Islam must be invited to embrace Islam before they may be legitimately attacked. Saudi cleric Sheikh Salman al-Oadah condemned bin Laden's and other al-Qaeda's methods as counter-productive and not in accordance with Islamic law. On the other hand, others have said that terrorist acts are inspired by the example of Prophet Mohammed's struggle in the just war against the Quraysh, the pagan tribesmen of Mecca.

Han Kung, a Catholic priest and President of the Foundation for a Global Ethic has concluded: "The apologetic argument often advanced by Muslims that armed jihad refers only to wars of defense cannot be maintained. It is contradicted by the testimonies of the Islamic chroniclers, who show that the jihad was of the utmost political and military significance. It is hard to imagine a more effective motivation for a war than a struggle… which furthers God's cause against the unbelievers" (The New York Review, Nov 8, 2007).

Revenge has always been a factor, for example, following the killing of a friend, family member or lover. Pathological depression or other forms of mental illness are not thought to be an important factor. As in military operations during wartime, bonding with one's comrades may be. Others have blamed poverty, Arab nationalism, alienation, charismatic shieks, religious fervor, dictatorial regimes, and feelings of inferiority. Many, if not most, jihadists act under the influence of psychotropic drugs. Many are drug dealers, who include explosives among their drug products.

11.13. Good vs. Evil

The new religion that appeared in Persia in the third century CE, called Manichaeism, taught that all matter, including the human body, was intrinsically evil. Other religions, including Christianity, teach that human beings are intrinsically good, and that sin is a failing to follow one's natural instincts to do good.

In his book Tree of Smoke, (Farrar, Straus and Grouse, 2007), Denis Johnson writes that there is a contest between good and evil in every human being. In war, soldiers are exposed to violence, the risk of death, the ecstasy of apostasy, voluntary degradation, and excesses. A sense of dread may become intoxicating. "Something wild, magical, stunning might come from the next moment, death itself might erupt from the fabric of this very breath, unmasked as a friend…"

In his book Democracy in America (1835–1840), Alexis De Tocqueville wrote "Satan, the enemy of the human race, finds his most powerful weapons in the ignorance of men, and it is important that the enlightenment brought here by our forefathers not remain buried in their graves…In America, it was religion that showed the way to enlightenment; it was respect for divine law that showed the way to freedom. Liberty looks upon religion as its comrade in battle and victory, as the cradle of its infancy and divine source of its rights. It regards religion as the safeguard of mores, and mores as the guarantee of law and surety for its own duration."

Historically, coupling religious convictions to politics has resulted in disastrous consequences, including pogroms, terrorism and war. Christianity was the dominant political and social force in Europe and other parts of the world for much of the last 2,000 years, with particularly tragic consequences for Jews. In the eighteenth century, the Enlightenment was characterized by faith in the triumph of universal human reason, benevolence, tolerance and pluralism. One of its core values was the separation of the church and state, was enshrined in the United States constitution. The religious orientation of the Enlightenment was a radical departure from Evangelicalism, which emphasized faith and Biblical truth (Head and Heart: American Christianities, Garry Wills, Penguin Press, 2007)

In his book The Varieties of Religious Experience (1902), William James concluded that "religion works," because it passes the pragmatist's test that opens a person to broader insights, greater generosity, self-sacrifice, greater usefulness to oneself and others, and opens new possibilities in life.

In their book Understanding Ethnic Conflict (2006), Raymond C. Taras and Raja Ganguly argue that military conflict among ethnic groups occurs when a nation collapses. In the anarchy that follows, ethnic groups fear for their survival. Violent conflicts result from the competition for power. Each ethnic group considers others to be the enemy. They use their wor apunis hir a ba the initiative. The orhustion in Iraq today is an example, as was the war that broke out between the Serbs and Croats following the collapse of Yugoslavia. Ethnic groups do not want to be assimilated, because they fear for their survival.

Accompanying the cultural, political and economic forces that lead to violence, biological factors are also involved. These include the use of illicit drugs. Trafficking is a major source of money and power for insurgency groups. The money is used to obtain weapons. Examples include illicit drug activities in Thailand, Laos, Burma, Turkey, Afghanistan and Iraq. Insurgents and drug dealers are everywhere in the Golden Triangle – Myanmar, Thailand, and Cambodia – and in the Golden Crescent – Iraq, Iran, Afghanistan and Pakistan – producing opium, hashish, heroin and cocaine.

Psychologist Edward O. Wilson believes that emotions and thoughts are strongly influenced by one's genes (Consilience, The Unity of Knowledge, 1998). He has "a conviction, far deeper than a mere working proposition, that the world is orderly and can be explained by a small number of natural laws." These include the "laws" related to brain chemistry, which is now the subject of research using molecular imaging.

Whenever new physical or chemical measurements involving the brain are developed, inevitably some try to use them as evidence of a lack of free will. Some lawyers defend their clients by denying their personal responsibility. The law assumes that every one has a free will, while science assumes that there is a cause for everything. The search for objective evidence of the brain processes that affect behavior goes on.

Violence and crime occur during stressful social, economic and political periods. When cultural forces are weak, human behavior seems to regress to primitive aggressiveness and violence. Social constraints are weakened. In experimental primates, sham rage can be produced by removal of the cerebral cortex. After decortications, the animals respond with fear and anger to trifling provocations. Norepinephrine can produce sham rage.

Sham rage produced by brain stem transection in cats is accompanied by changes in brain stem norepinephrine. Rage is increased by protriptyline that increases the action of norepinephrine, and by haloperidol that depresses its action.

Paul Soloff and colleagues from the University of Pittsburgh have found prefrontal hypoperfusion and decreased glucose uptake in the prefrontal cortex (PFC) in violent criminal offenders, murderers and aggressive psychiatric patients. FDG-PET studies were also performed after a fenfluramine challenge. In healthy volunteers, fenfluramine led to increases in glucose utilization in the left prefrontal and left temperoparietal cortex. Studies were performed in persons with borderline personality disorder (BPD), a serious mental illness characterized by pervasive instability in moods, interpersonal relationships, self-image, and behavior. In these persons, fenfluramine produced bilateral frontal cortex activation and decreased activity of the temporal cortex and thalamus.

Siever and colleagues studied six impulsive-aggressive patients and five healthy volunteers to determine their response to fenfluramine compared to a placebo. Normal volunteers showed increases in the orbital frontal and adjacent medial frontal cortex, cingulate, and inferior parietal cortex. Patients with impulsive aggression showed significantly blunted metabolic responses in the orbital frontal, adjacent ventral medial, and cingulate cortex, but not in the inferior parietal lobe, when compared with their matched controls. Significant reductions in FDG uptake were found in persons with BPD compared to healthy controls. The reductions were found bilaterally in the medial orbital frontal cortex, including Brodmann's areas 9, 10 and 11.

"The relentlessness of modern science's progress, which constantly corrects itself by discarding the answers and reformulating the questions, does not contradict science's basic goal – to see and to know the world as it is given to the senses – and its concept of truth is derived from the common-sense experience of irrefutable evidence, which dispels error and illusion" (Hannah Arendt. *The Life of the Mind*, Harcourt Brace Jovanovich, 1971).

The human race today faces extreme violence. The Buddhist philosophy teaches us that nature is not intrinsically good, that pain and suffering are the way the world is, that we must live with the inevitable imperfections of life. The Judeo-Christian tradition teaches that everything happens for the best. The pragmatist view is that there is no absolute truth, that both of these ideas must be taken seriously. William James tells us that: "The true is the name of whatever proves itself to be good in the way of belief, and good, too, for definite assignable reasons." Success is not assured, but perhaps taking little steps toward understanding the chemistry of the brain can offer some reassurance that the human race can survive. "The world stands really malleable, waiting to receive its final touches at our hands." (William James, *Pragmatism. New Impression*, Longmans, Green and Co. NY, 1914.)

Knowledge begins with sensory impressions. When we look at a star, we integrate the image with our memory of the properties of a star. When we say we hear a bell, what we actually hear is the sound of a bell. What we see or hear is not the object or the sound, but with the effect that the object or sound has on us. Knowledge and ideas are derived from perceptions. Sensations, perceptions and thoughts lead to actions.

Physical and chemical events in the sensory organs and brain are responsible for sensations and perceptions. Neurons convey facts from the sensory organs to regions of the brain involved in perceiving. Our perception of a star represents a star that does not exist at the time we see it, because the source of the light has burned out by the time the light reaches us. A rainbow does not have an objective existence, but is a perception. When a psychiatrist showed Rorschach pictures to his patient, every response had sexual connotations. The psychiatrist asked: "Don't you ever think of anything but sex?" The patient answered: "You're the one showing all the dirty pictures."

We interpret perceptions in the "light of knowledge stored in our brains. When we see a person that we haven't seen for years, we say: "You haven't changed a bit." We have stored several of the person's essential characteristics in our brain, rather than the totality of his or her existential appearance. Years ago these features characterized the person. These have not changed over time. On the other hand, if you compare the person's face to a photograph taken years ago, the changes over time are dramatic because we are comparing thousands of the person's present features with the photograph. When we look at the person later in real time, we relate his or her features to images that we have stored in our brain. Properties from past experiences that represent color, shape and other features are stored separately in different parts of the brain and later integrated into mental images. Facial recognition involves the amygdala and orbitofrontal cortex.

How are sensory data about objects recorded in the brain? Do certain molecules in the brain record the shape, color, and other features of the objects that we sense? Do chemical processes that encode the characteristics of the material objects that we sense? Is there a "perception code" in the brain analogous to the code of ATCG nuclides that provide the basis of the genetic code? What are the substances within the brain that provide such a "perception code?" How do these substances make up a code? Is it possible that the same principle of a four-letter code contained within each neuron is the brain of the "perception code"? Is it possible that three dimensional objects in real space are encoded in the patterns in the code as represented in the spatial configuration of the neurons that encode the objects? Memory involves amino acids, such as glutamic acid, which is involved in bringing sensory information to the brain, and in rearranging synaptic connections related to memory. Removal of a chemical group results in neuronal-stimulating glutamic acid becoming a neuronal-inhibitor, gamma aminobutyric acid (GABA).

Life is a continuing struggle against entropy. The neuronal and chemical activities in our brain follow the laws of physics and chemistry. "Consciousness" is the emerging result of on-going electrochemical processes in the brain in response to what is happening in the environment, integrated with memories of past perceptions and ideas.

C. Roy and Charles Sherrington were the first to describe the close relationship between regional blood flow and neuronal activity (J. Physiol 11, 85–108, 1890). Beginning in the early 1990s, an MRI technique called *Bold MRI* measured regional hemoglobin oxygenation as an indicator of regional increases in neuronal activity. This method has far greater spatial resolution than PET studies with oxygen-15 water.

The mind emerges from the electrical and chemical activities of the brain, just as story emerges from a book. The mind is what the brain does, just as walking is what the legs do, and seeing is what the eyes do. It emerges from the billions of neurons in the brain, each neuron being connected to other neurons by tens of thousands of synapses. William James proposed "the thoughts themselves are the thinkers."

Consciousness is a construct. For example, the world doesn't contain colors. We perceive colors because matter reflects electromagnetic energy of different waves lengths from the sun. Our eyes and brain convert this energy to the consciousness of color. The brain also makes us conscious of smells. Without the brain, there would be no red color or sweet smells. As we view events in the environment, our eyes flit from place to place, focusing on what attracts our attention. Perception involves countless numbers of neurons. Perceptions are integrated with the knowledge provided by neurons that are active in working memory. Information perceived from an enormous amount of sensory data from the past is integrated with short-term memory to provide a "feeling of self." For example, visual cortical area IV is activated when a person views a scene or when he only imagines it (Stephen, 1994).

Should morality determine our actions? In 1784, Maximilien Robespiere wrote: "Let us establish morality upon an eternal and sacred basis; let us inspire in man that religious respect for man – the profound sentiment of his duties, which is the sole quantity of social happiness."

In the New York Times on June 22, 2007, David Brooks wrote that "preaching morality to students doesn't change behavior… Schools are ineffective when it comes to values education…We're primarily perceivers. The body receives huge amounts of information from the world, and we turn that data into a series of generalizations, stereotypes and theories that we can use to navigate our way through life. Once we've perceived a situation and construed it so that it fits with one of the patterns we carry in our memory, we've pretty much rigged how we're going to react, even though we haven't consciously sat down to make a decision. Construing is deciding...Deciding is conscious and individual, but perceiving is subconscious and communal...We're perceivers first, not deciders."

Our brains would be swamped by the enormous amount of information continuously reaching them without this selection process in our working memory. We direct our attention, perceive, and integrate the information with our working memory, and then act. There is no *homunculus* in the brain.

Molecular imaging provides a way to study of the relationship between the brain, the mind, and behavior. With radiolabeling drugs that bind to receptors, we can relate the regional chemical processes to the observed effects of the drugs. It may be possible to relate chemical processes in different parts of the brain as we perceive the outside world and decide how to act. We want to know the correlation of thoughts and feelings with molecular processes in different parts of the brain. For example, we know a lot about the chemistry of human vision. These chemical processes do not cause vision, but are "manifestations" of the chemistry of vision, measurement of which may prove useful in solving a patient's medical problems. Similarly, we don't know what causes gravity; but we live with its manifestations. We don't know why the speed of light is constant, but we used the information to make atomic bombs.

Our consciousness can be expressed in words. Language is mankind's greatest invention, even if communication is our greatest problem. Words reveal to others our perception of reality.

"Reality is nothing but a text that we write and continually rewrite" (Jacques Derrida).

"What gives the world around us form and substance is our contribution...The universe is our own making" (Frayn, 2007). Consciousness of thoughts, emotions and desires creates our personal world.

The bumping and jostling of atoms and molecules in our bodies increase entropy. Survival depends on the continual supply of energy provided by the mitochondria in the nucleated cells of our bodies, which must be obtained from what we eat and drink. The complexity of living organisms involves an enormous amount of information, culminating in the complexity of the human body. New information and new complexity are the hallmarks of evolving life. From the time we are born until we die, we depend on the communication systems needed to maintain such a complex system, involving sensory information, language, and autonomic functions that keep us alive. We can now examine the chemistry of communications within the brain and other organs of the body. All human thought, hope, fear, passion, and insight result from chemical interactions between neurotransmitters and receptors.

11.14. Aging

For a long time, it was thought that the brain cannot generate new neurons throughout adult life; but can only create new synapses. New evidence shows that even later life, the brain may continue producing neurons. We do lose some neurons as we grow older, but the normal aging process leaves most mental functions intact. Studies with PET/FDG suggest that overall brain activity does not decrease significantly with age, although certain regions have less activity.

Marilyn Albert of Johns Hopkins (Wall Street Journal, March 3, 2006) said: "It used to be thought that normal cognitive decline occurred because of the loss of neurons throughout the brain." Now it is thought that most regions of the brain hold on to their neurons, and that even 70-year-olds produce new neurons, with little or no loss in the hippocampus which is the control. In the frontal cortex, the site of executive functions, such as planning and judgment. The elderly tend to have fewer new experiences, are less active physically and socially, and live in less complex surroundings, which may impair the maintenance of neural circuitry.

"The trouble with retirement is that there are not a lot of social or intellectual demands," says research psychologist Denise Park of the University of Illinois. "Life becomes routine, a recipe for cognitive decline." Some of the decline attributed to aging does not reflect aging per se but factors under the aging person's control. Attention and focus are controlled by the prefrontal cortex. Older persons use both the right and left half of the brain to carry out functions that the young accomplish with only one side of the brain. Older adults who activate both the left and right prefrontal regions have good short-term memory. The use of both sides of the prefrontal cortex compensates for deficits in the hippocampus.

K.I. Erickson, and AF Kramer (*Brain and Cognition,* 2002) believe that old brains can be trained to act like young ones, and that "...brains of older adults remain relatively flexible and able to alter brain circuits in response to training."

Forty years ago, Joel Elkes wrote: "It is encouraging that methods for the analysis of regional transactions within the brain are now becoming available. The quantitative histochemistry of some enzyme systems in relation to cell populations is becoming a feasible undertaking, and is bound, sooner or later, to be related to the action of drugs at subcellular sites...It is the interaction between regional neurochemistry, electrophysiology, mathematics and behavior which holds out the highest promise for a coherent science of behavioral pharmacology of the future." Today, imaging of brain chemistry has begun to play a major role in brain science.

Elkes believed that progress in technology would let us move steadily ahead, becoming "better and better" and "truer and truer." Forty years later, the New England Journal of Medicine would designate medical imaging as one of the 11 most important innovations of the past 1,000 years. Molecular imaging with PET and SPECT lets us progress deeper and deeper into the domains of the cells, genes, molecules, atoms, and sub-atomic particles in the brain, learning more and more about how the brain works and how to make it work better. Thinking guides us to meaning; knowledge guides us to truth, as the physician tries to answer eight questions:

1. What is the patient's problem?
2. Where is the problem?
3. What is going to happen?
4. Can anything be done about it?
5. What is the best choice of treatment?
6. Is the treatment helping?
7. Is there still a problem?
8. Has the problem recurred?

Molecular imaging with PET and SPECT, and other imaging technologies, including CT, MRI, and optical imaging help answer these questions. PET imaging with F-18 FDG has been used extensively in patients with stroke, epilepsy, Alzheimer's disease, Parkinson's disease, and Huntington's disease; and in psychiatric disorders such as schizophrenia, depression, obsessive-compulsive disorder, attention deficit hyperactivity disorder (ADHD), and Tourette syndrome. PET can distinguish patients with depression from those with dementia, and can characterize the extent of brain tumors. It can localize the site partial complex seizures, localize the site and size of strokes, and monitor drug treatment of patients with mental disorders.

Specific neurotransmitters carry hundreds of different messages. Each neuron is affected by thousands of different neurotransmitters that cross synapses to be received by receptor sites. Neurotransmitters can be thought of as intelligence floating along the rivers of the brain, bumping, fitting, and sticking to the receptors lining the banks. The brain creates the mind from enormous amounts of chemical information.

Genetic instructions govern the expression of specific molecules that control huge networks of individual neurons. Every neuron has the addresses of hundreds of other neurons from which it receives and transmits molecular messages through its axonal terminals and synapses. Synaptic growth depends on genes that encode interneuronal molecular signals. A neuron deprived of presynaptic molecular messages whithers and dies.

One group of chemicals, called endorphins, increases tolerance to pain, acting like morphine. Other neurotransmitters, including norepinephrine, cholecystokinin (CCK) serotonin and dopamine are also involved. Unpleasant perceptions, pain and addiction involve similar molecular processes.

Dopamine is present in high levels in the brain of patients with schizophrenia. Their treatment is based on blocking dopamine receptors. Chemicals regulating our emotions are rich in the amygdala and hypothalamus, but are present in great abundance outside of the brain as well, for example in the monocytes of the immune system. The latter can be thought of as circulating neurons. DNA is the chemical mastermind in all of our cells (except red blood cells).

Disease can be defined according to the sites and types of specific biochemical processes. The disease originally called Graves' Disease and now called hyperthyroidism is manifest by increased expression of the sodium/iodide transporter, which results in increased accumulation of iodide, increased synthesis of thyroxin and other thyroid hormones, and increased cellular metabolism. Detection of one or more of these abnormal processes can be *holons* in the diagnosis of a patient's disease.

The body, including the brain, involves hundreds of thousands of chemical processes. We need to arrange these to create a systematic classification of diseases. Such a system is made up of holons that are part of a *holonistic* network. Identification of "sub-holons" that are abnormal will make possible the systematic analysis of the manifestations of the specific patient's disease. The diagnostic process will proceed "upward" from progressive sub-assemblies to progressively more complex, or "downward" from the abnormal anatomical structures or physiologic processes. As Arthur Koestler has said: "Wherever there is life, it must be hierarchically organized."

We can image the spatial and temporal distribution of thousands of molecules within the living body by means of instruments outside the body. Molecular processes in different parts of the body will be the basis for a whole new definition of disease. There will be new ways of defining disease, which eventually will become accepted and used in ways not dreamed of today. Long after an important happening has occurred, people do not say: "I was for it or I was against it," but rather "I didn't know it was happening."

The fifth edition of the Diagnostic and Statistical Manual of Mental Disorders (D.S.M) is scheduled for print and for 2012. The first edition was in 1952. It categorized mental illnesses on the basis of symptoms. According to Sally Satel (NY Times, Sept 13, 2007), "three decades later, psychiatry still lacks a firm grasp of the causal underpinnings of mental illness...Many patients meet several diagnostic definitions at once." Today, we search only for causes of mental illness, but for "modifiable molecular manifestations (3M) of mental dysfunction."

When a person feels ill, and decides to "see a doctor," the challenge for the doctor is to "make the diagnosis," that is, to identify the disease from which the patient is suffering. There is the basic assumption that the patient is not suffering from a unique illness, but one similar to that of others. The patient's illness may not be exactly the same as that of any other person, but there are enough similarities to make it possible to use knowledge of similar patients to help care for the patient in front of him.

Long ago, and to some extent today, diseases were defined by symptoms, such as headache, chest pain or cough. The diagnosis might consist of translating the patient's symptoms into Latin. For example, dolorosa rubra nodularis was the diagnosis in those patients who had painful, red nodules in the skin.

In 1769, the Italian physician Giovanni Batista Morgagni published: *The seats and causes of diseases investigated by autopsy*. Autopsies and histopathology became the foundation of scientific medicine. The challenge to clinicians was to correctly predict during the patient's life, what would be found at autopsy. The clinical diagnosis was made by the process of *differential diagnosis*, the listing of all possible diagnoses, consistent with the clinical and laboratory findings. Further tests were performed to confirm or refute each of the various diagnostic possibilities.

Karl Popper said it is impossible to confirm a hypothesis. It is only possible to refute it. He rejected the probabilistic approach to medical diagnosis, which today is the most widely accepted model for the diagnostic process. He maintained that probabilistic induction is impossible most of the time because confirmation is impossible. This is not the case in medical practice, where biopsy and surgery can confirm a prior diagnosis. Popper stressed the importance of tests designed to "falsify" a hypothesis. This applies to the probabilistic approach to the diagnostic process.

According to the ontological viewpoint of disease, diagnoses should be expressed together with statements of the probability of each disease that may be present. A mathematical statement of the probability for each possible disease reflects the certainty of each possible diagnosis being correct. For example, the physician may conclude that the patient has a 90% probability of suffering from Alzheimer's disease, manifest by progressive loss of memory. Other possible diagnoses might include mild cognitive impairment (MCI), depression, Parkinson's disease or an endocrine abnormality, such as hypothyroidism. Further testing progressively eliminates possible diagnoses. The tests also serve as guides to the care of the patient. "Working diagnoses" may be born out by further tests or by following the course of the patient's illness.

The effective practice of medicine depends on the physician's knowledge, experience and diligence. Definitions of disease are essential to the diagnostic process. An important part of a doctor's education is learning how to obtain and use knowledge of diseases. A challenge is to systematize the diagnosis of disease by molecular imaging as it becomes used more and more in medical diagnosis.

In 1703, Gottfried von Leibniz wrote: "Nature has established patterns originating in the return of events, but only for the most part." This is a key principle of the diagnostic process. It is "probabilistic," that is, the patient's diagnosis is always uncertain, because no two patients are exactly the same, and what happens to a specific patient may not be the same as what happened to all the patients treated for the same disease. Actions must be definite even in the presence of uncertainty.

Galileo said: "The book of nature should be written in the language of mathematics." Molecular imaging of biochemical processes makes it possible to establish mathematical criteria to help predict what is going to happen to a patient who manifests mathematically defined abnormalities in certain regions of the body. Mathematical criteria are used to decide what should be done to determine whether the treatment is working.

Steps that a doctor takes in making a diagnosis haven't changed. Diagnoses are hypotheses. The doctor asks: "What's wrong? The answer is called the patient's "chief complaint," followed by a description of the patient's "past history" and "present illness," going back to the time when the patient last felt well.

When you were a child, you may have been diagnosed as "having a strep throat." In your later school days, you learned that you had a "sprained ankle," fortunately not a fracture. When you reached middle age, you were found to have "high blood pressure," arthritis, or, sad to say, early Parkinson's disease. You were glad that you didn't have "Alzheimer's disease." The "differential diagnosis" was the list in your doctor's mind of the diseases that you might have. These had to be differentiated, the correct diagnosis made and treatment started. Making the diagnosis was the first step.

Simply talking to the patient may suggest the diagnosis. The next step is the "physical examination," performed with special instruments, such as an ophthalmoscope or stethoscope, to detect physical signs characteristic of specific diseases. The examination would next include obtaining blood and urine samples.

Doctors obtain information to try to answer four basic questions that make up the practice of medicine: what is wrong? What is going to happen? What can be done about it? How did it happen?

Søren Kierkegaard (1813–1855) said: "We live forward but can only think backward." For over 100 years, a disease has been defined as a histopathological abnormality of an organ or tissue, caused by an infection, genetic defect, or environmental stress, and characterized by an identifiable group of clinical symptoms and physical signs.

Diseases often have characteristic symptoms, such as headache; or physical signs, such as the paralysis of facial muscles; or by abnormalities found at autopsy, such as a brain tumor; or by histopathological abnormalities in tissues and cells, such as inflammatory cells. Diseases are also detected by abnormal blood tests, such as an elevated blood sugar level, or as regional abnormalities in molecular processes. Half a century ago, the focus of much medical research was to detect abnormalities in the blood of molecules, such as insulin, ACTH, hydrocortisone, thyroxine and other hormones. It was not possible to examine the chemistry of the brain, except at autopsy or when brain tissue was removed at surgery.

Four categories of diagnostic information are: clinical manifestations, structural abnormalities, histopathological abnormalities, or abnormal molecular processes.

Collections of characteristic manifestations of disease are called syndromes, which may not include a proven pathological or pathophysiological association. However, most diseases are defined by specific pathological or pathophysiological manifestations. Definitions of many diseases are vague, especially mental diseases, including schizophrenia, anxiety and depression.

It is important to consider the patient as a person. Increasingly, this is being called "personalized" medicine. Patients are characterized by specific, quantifiable molecular manifestations, progressing from genotype to phenotype. Just as human beings are not all the same, neither are diseases all the same. Every patient is diversely different. Each patient has unique features that are essential for optimum solution of his or her problems. It is often not possible to describe a patient with only one abnormal manifestation of disease, or as having a single disease.

Accurate communication is essential when physicians call upon consultants. Evidence used in the interpretations of results by referring physicians should be included in the report to the primary physician, and must be clear when promptly transmitted to the referring physician, and preserved as the patient's medical record. The report should include the images on which the verbal interpretation is based, as well as probability statements regarding the abnormalities or whether the study reveals no abnormalities.

The report of imaging studies should include (1) Priori clinical and other data; (2) The patient's principal problems; (3) Results of the imaging procedures; (4) Location(s) of anomalies in the distribution of the tracer; (5) Characteristics of the abnormal tracer distributions; (6) Quantification of regional accumulation sites; and (7) Interpretation of the study in the light of the patient's problem.

In the late 1950s, I began to use Bayes' theorem in the interpretation of imaging studies. Its basic premise is that the diagnostic process is probabilistic and should be based on the combining of the prevalence of specific diseases in the population from which the patient comes and the specificity of the objective findings in the patient being diagnosed.

In the ontological approach to medical diagnosis, each disease is thought of as a distinct entity. An alternative approach to the diagnostic process is the physiological approach. Disease is defined in terms of objective manifestations. For example, the disease *hypertension* is based on a physiological measurement. Hyperthyroidism is characterized by an increased rate of accumulation of radioiodine by the thyroid gland. Different diseases may have some of the same manifestations. Examples are senile dementia of the Alzheimer type (SDAT) and idiopathic Parkinson's disease.

The diagnostic process is probabilistic. Present day definitions of disease are often crude and inconsistent, which is why quantifiable criteria are helpful. Histopathological criteria for classification suffer from being subjective in their interpretation, and are based on the study of dead tissue, even when it has been obtained by biopsy. The befuddling influence of a specific diagnosis used as the sole basis for defining the patient's problem results from the belief that diseases are unalterably divided. In fact, it is usually hard to characterize a patient by stating only the "disease that the patient has." It better to say that the patient's blood pressure is 230/100, or simply that the patient has hypertension? Many diseases are linked, for example, Alzheimer's and Parkinson's disease (vide infra). Patients often have both motor and cognitive deficiencies. It is not helpful to force a patient's disease into one diagnostic category. For example, in making the diagnosis of Parkinson's disease, it is helpful to measure the degree of dysfunction of D2 dopamine receptors or of the presynaptic dopamine transporters. These measurements influence therapeutic decisions.

Radiotracers make it possible to measure many molecular processes within different regions of the body. While it is tempting and helpful to define diseases only in terms of objective measurements, there is almost always a subjective component. For example, the information in molecular medicine is almost limitless, only a small part can be quantified.

Questions can be answered objectively with quantitative data that can direct specific treatment and monitoring of its effectiveness. While focusing on the information relative to the patient's problem, one can at times detect unanticipated abnormal findings, some of which could be more important that the quantifiable data. Every fact revealed by the molecular images needs to be detected and examine, even if originally thought to be incidental to the patient's chief problem(s). In terms of a legal analogy, every suspicious observation must be acquitted or followed up.

For example, if a patient has a PET scan with F-18 FDG because of the finding of a nodule in the lung revealed by a chest x-ray, the interpretation should involve quantification of the rate of accumulation of the tracer by the lesion, together with a probabilistic statement of the degree of its malignancy and the presence of lesions elsewhere in the lungs or whole body. In view of the uncertainty in medicine, statements should whenever possible involve probability of certainty statements.

All data, both objective and subjective, should be used in the interpretation of the results. The first interpretation of the images should be completely objective, without consideration of any a priori information, followed by a second interpretation based on the imaging results in combination with all in the light of all of the data.

Many patients have problems that cannot be defined quantitatively, and decisions must be made with approximate information and uncertainty. There are formalized tools for dealing with the imprecision intrinsic to many problems.

In 1960, Lotfi Zadeh proposed what he called "Fuzzy Logic" to describe the way people think and speak in their natural language. *Fuzzy set theory* creates classes or groupings of data with boundaries not sharply defined. *Fuzzy* techniques are used to solve real-world problems where we must deal with imprecision in the variables and parameters that are measured

Fuzzy set theory encompasses fuzzy logic, fuzzy arithmetic, fuzzy mathematical programming, fuzzy topology, fuzzy graph theory, and fuzzy data analysis, though the term fuzzy logic is often used to describe all of these. Fuzzy logic often is based on terms such a "large," "medium," and "small," rather numerical values.

Fuzzy logic allows overlap of values (e.g., a 85 kg man may be classified in both the "large" and "medium" categories, with different probabilities of belonging to each category).

Fuzzy logic moved into the mainstream of information technology in the late 1980s and early 1990s, as a departure from classical Boolean logic in that it implements soft linguistic variables on a continuous range of values. Fuzzy logic is used to model complex systems where ambiguity or vagueness are common.

In summary, we not only search for the causes of the patient's problem, but also search for modifiable molecular manifestations, their location, structural abnormalities,

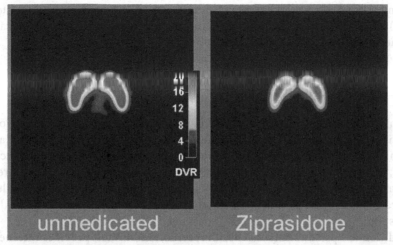

Fig.11.1 C-11 raclopride images of a patient with schizophrenia before and after treatment with Ziprasidone, which binds to dopamine, serotoninand alp ha-adrenergicr eceptorsi nt heb asalg anglia.

function and regional biochemistry. Eventually, the patient's diagnosis will be his or her name, characterized by the manifestations of the problem. Using molecular imaging, research will be directed to find the location and nature of the abnormal molecular processes that can be restored to normal. The new language of pathophysiology is regional biochemistry. For example, fear and anxiety will be characterized by chemical abnormalities in the amygdala of the brain.

11.15. Information Technology

Data obtained by imaging must always be related to clinical information, as well as other data, such as histopathology. Referring physicians must be consulted to view and discuss their patients' studies. High quality images must be available to referring physicians and other doctors responsible for patients' care. Fusion of PET with CT and MRI images makes molecular imaging more understandable to referring physicians and patients, and enhances collaboration with referring physicians, better interpretation of the studies, and more efficient flow of patients. Consultations and second opinions are often necessary.

Information technology is not an enormously complex and expensive venture. Distant viewing of images with 512 × 512 picture elements (pixels) must be available. Digital images can be transmitted without loss of fidelity over transmission lines at low cost. A modern laptop computer can easily download PET, SPECT, CT and MRI or fused images, and the data can be fed directly into an Internet server. (The main obstacle in putting information technology into medical imaging is hesitancy to accept new things. When we first introduced computers into nuclear medicine in the late 1960s, in the beginning our trainees would hesitate to enter the rooms where the first computers were located.) The Internet can provide vastly more information and greater logical ability than that of any single human being can possess.

11.16. Electronic Health Records

Electronic health records (EHR's) are key components of our "Information Age," putting information about the patient's health at the physician's fingertips at all times and in all places. Microsoft now makes it possible to put personal health records on the Web, in a program called *HealthVault*. Partners include hospitals, disease-prevention organizations and health care institutions. For example, in Baltimore, Health Vault collaborates with Union Memorial Hospital, Harbor, Franklin Square and Good Samaritan Hospital.

In addition to a person's personal health records, Health Vault provides Internet search capabilities. The data are stored in an encrypted database. Each person will select which of their personal health care information is put into the Web database. Most persons will give permission to their doctors, clinics and hospitals to send directly to HealthVault, information concerning medicines, test results and other data. At New York Presbyterian Hospital, Aurelia G. Boyer, the chief information officer, states their commitment to helping patients manage their own health care.

The Patient Privacy Rights Foundation has praised Microsoft's privacy policy that allows users to control which information they provide through an opt-in program. In its privacy page, Microsoft says "We do not use your health information for commercial purposes unless we ask and you clearly tell us we may."

An example of an MRI and PET database is the Alzheimer's Disease Neuroimaging Initiative (ADNI), launched in 2005 funded by $60 million over a 5-year period. To date, 800 patients have been enrolled (www.loni.ucla.edu/ADNI).

Information Technology (IT) has been called a *disruptive technology*, not because it interferes with the traditional personal doctor/patient relationship, but because it will leave behind those who do not embrace it. The invention of the "electrical speech machine" by Alexander Graham Bell was a disruptive technology that revolutionized communication. Information ethnology helps standardize and speed up medical care so that more patients can be treated at reduced cost. IT extends the physician's brain by providing information at his or her fingertips.

For more than 20 years at the Regenstrief Institute at the University of Indiana, Clement McDonald has been using a medical record system that now holds more than 480 million distinct laboratory and clinical measurements, 40 million radiology images, and a variety of other kinds of data. Since 1994, five major hospital systems in Indianapolis participate in this program along with public health departments, physician offices, laboratories, and other care providers.

Every referring physician with a computer, browser and viewing software (and the appropriate password and decryption key) can access and view the studies of his or her patients. In home workstations, a storage capability is not needed, because studies can be retrieved from the server. Only four abilities are needed: availability, capability, affability and marketability. Here is how the system operates. Today, every morning the medical imaging specialist calls up a list of the patients scheduled for imaging studies that day. A click of a mouse will bring up their electronic health records (EHR's). In interpreting the studies, if needed, a quick search of the literature can provide answers to questions that arise. Programs can also be called up to see if there are patients with similar problems in the databanks.

Hand-held computers are becoming commonplace when a physician is talking to patients. When I was an intern, I was concerned initially that my typing while taking the patient's history might interfere with my relationship with the patient. This did not turn out to be the case. Rather it indicated to the patient my high level of concern. From the early days of my training in internal medicine, I learned the importance of going to the Radiology Department every evening to consult with radiologists about X-rays performed that day. Persons with the ability and dedication of Drs. David Gould and Dan Torrance were always willing to educate us young, inexperienced physicians in the process of telling us the results of our patients' studies. Today, immediately or soon after an imaging study is completed, the processed images should be viewed by the imaging specialist together with the referring physician via the Internet. In interpreting the studies, experts in other specialties will be contacted immediately when necessary.

We can now communicate instantaneously, anywhere, anytime. Patients can be connected to their physicians via the Internet. For example, after a hospital visit, a patient wants to know as soon as possible: "Do I have cancer? How long has it been present? Can it be cured by surgery? Can I make it through the operation? How long do I have to live?"

Information technology makes it possible for both physicians and patients to keep up with new happenings in clinical medicine, including technological advances, such as molecular imaging, in solving present and future problems.

Nuclear medicine physicians have been leaders in the use of medical information and communication systems in their daily practice. They know how important it is to keep up-to-date on new technology, as it develops. Information technology is transforming health care, improving the personal relationships essential for the practice of medicine. The practice of medicine is changing as never before. Information technology is bringing patients and physicians closer than ever before, helping to provide faster, better and cheaper care of our patients.

11.17. Personal Health Chips

Rather than using a commercial Web connection, such as Health Vault, some persons may prefer to carry a periodically updated digital computer chip that contains the lifetime manifestations of their health.

The screening of images stored on their chips can provide a baseline making possible the detecting of disease at the earliest possible time. Abnormal manifestations of disease will be identified, and then personal physicians will interpret the abnormal findings, and provide medical advice and treatment.

The person's health chip will permit periodic searching of an international health manifestation database (IHMD) to identify emerging health problems, predict outcomes, identify new treatments, evaluate ongoing therapy, and predict outcomes. The data from scientific meetings and the medical literature will be continually examined to keep the IHMD current. Physician reviewers will identify and format new medical literature. The patient's anatomical, functional, and molecular imaging studies will be compared with images in the IHMD using co-registration of the images. Existing search engines can scan six billion sites in 90 languages for 138,000 people every minute.

12

Drug Design and Development

Positron emission tomography (PET) and single photon emission computed tomography (SPECT) are now being used widely used in drug design and development to characterize specific molecular abnormalities in different regions of the brain in patients with stroke, epilepsy, Alzheimer's disease, Parkinson's disease, and Huntington's disease, as well as psychiatric disorders, such as schizophrenia, depression, obsessive-compulsive disorder, attention-deficit/hyperactivity disorder, and Tourette syndrome.

Molecular imaging has joined electroencephalography (EEG), magnetoencephalography, functional magnetic resonance imaging (fMRI), and diffuse optical tomography in measuring neuronal activity. fMRI has a spatial resolution of 1 mm^3 and excellent temporal resolution. Both PET and SPECT have spatial resolution of about 1 cm^3, but low temporal resolution.

fMRI is based on measurement of changes in blood flow and blood oxygenation in the regions of the brain that have increased neuronal activity. Regions of the brain activated when subjects perform a task–the "brain in action" can be identified by imaging the regional increases in blood flow and oxygen consumption with fMRI in different regions of the brain. The high spatial resolution makes it possible to detect changes in 2 or 3 mm of brain, which includes thousands of neurons.

When neurons are activated, the neurons consume more oxygen from the hemoglobin in red blood cells. The increases in blood flow associated with increased neuronal activity begin after 1–5 seconds, and rise to a peak in 4–5 seconds.

fMRI, PET, and SPECT make possible visualization and quantification of the biochemical processes in the brain of living humans. In PET, radioactive tracers decay by emitting positrons that travel a few millimeters in the brain until they react with an electron, which results in the annihilation of both the positrons and electrons with consequent emission of gamma ray photons that travel through the brain in opposite directions. The emitted photons are detected by radiation detectors in a ring surrounding the patient's head to provide images of the distribution of the radiotracer within the brain.

SPECT uses radiotracers that emit single photons rather than positrons. Their distribution in the brain is measured by a ring of radiation detectors surrounding the patient's head, as is the case in PET imaging. SPECT can measure regional brain blood flow to reflect increased neuronal activity. PET and SPECT can image regional molecular processes, for example, the location and availability of neuroreceptors.

Specific neuroreceptors are identified by specific radiotracers injected intravenously. PET and SPECT images are often fused with computed tomography (CT) and magnetic resonance images (MRI) to show the detailed anatomic structures within the brain where the biochemical processes are occurring.

SPECT uses radionuclides, such as technetium-99m, rather than positron-emitters, such as carbon-11 or fluorine-18, as is the case with PET. A widely used SPECT radiopharmaceutical is technetium-99m hexamethylpropylene amine oxime, to measure regional cerebral blood flow after systemic administration of the cerebral vasodilator acetazolamide (acetazolamide test) to patients with Alzheimer's disease (AD) or vascular dementia (VD).

Technetium-99m and other metals can be incorporated into "coordination complexes," such as the N_2S_2, tricarbonyl, or S-peptide complexes, such as lanthionine, which are used to label amino acids, peptides, proteins, or other molecular recognition sites in the body. Other useful metals in addition to technetium are rhenium and gallium. An advantage of single photon tracers is that multiple radionuclides can be measured simultaneously, being separated by gamma spectrometry. The search for molecular manifestations of a disease such as Alzheimer's may take a top-down approach that involves both PET and SPECT, or a bottom-up genetic approach.

Some radiolabeled molecules target RNA and DNA, as well as proteins. Reporter genes also can be imaged. These are genes that can be attached to another gene of interest, because they are easily identified and measured as markers. Reporter genes are often used to determine whether a gene of interest has been expressed in a cell or organism.

Molecular imaging, genetics and pharmacology are the three legs on which drug design and development stand. The digital computer can create precise three-dimensional models of complex molecules, including hormones, enzymes, and genetic materials. This makes it possible to design drugs targeted to specific molecular sites of action with fewer side effects.

Many drugs bind to the molecular receptors that lie in the outer membranes of each cell along the banks of the "stream of life." These receptors lie in wait for molecules with the specific configuration that cause them to bind. As these molecules travel throughout the internal environment, described by Claude Bernard, they bump, fit and stick to the receptors. The receptors are analogous to keyholes into which only specific molecules fit. Drugs bump, fit, and stick to specific enzymes and other biomolecules in the symphony of life that controls our interactions with the environment.

According to Deepak Chopra (*Quantum Healing: Exploring the Frontiers of Mind/Body Medicine*, Bantam Books, 1989), "… the living body is the best pharmacy ever devised. It produces diuretics, painkillers, tranquilizers, sleeping pills, antibiotics, and indeed everything manufactured by the drug companies, but it makes them much, much better. The dosage is always right and given on time; side effects are minimal or nonexistent; and the directions for using the drug are included in the drug itself, as part of its built-in intelligence … Every cell in the body is programmed by its DNA … Everything is regulated by our inner intelligence … the river is constantly being changed by the new water rushing in … All of us are much more like a river than anything frozen in time and space."

The first step after identifying a biomolecule of interest is to radiolabel it with a positron–or single photon-emitting radiotracer. After toxicity testing in animals, it is injected first into animals and then into human beings in processes called pharmacokinetics (PK) and pharmacodynamics (PD). PD is the study of the biochemical and physiological effects of drugs and the relationship between drug concentration and effect. Drug–receptor interaction models include the ligand-L (drug), its receptor-R in a mathematical expression of their interactions.

PD is the study of what a drug does to the body, whereas PK is the study of what the body does to the drug. PD is often combined with PK, and is denoted PKPD.

The sites of accumulation of the drug in various organs and tissues can be imaged and quantified. Many radiotracers are ligands that bind to specific neuroreceptor subtypes, such as dopamine D_2 receptors, serotonin (5-HT)1A or enzyme substrates (e.g., 6-FDOPA). Occupancy of receptors by endogenous molecules, such as neurotransmitters, can be determined by their effect in decreased binding of competing radiolabeled compounds.

Molecular imaging with PET and SPECT is quicker and more cost-effective than the older technique of killing and dissecting animals. PET scanners specifically for the study of small animals are being widely used in the pharmaceutical industry, and increasingly in academia.

Biomarkers used in drug design and development include (1) receptors, (2) enzymes, (3) protein and peptide transporters, and (4) intracellular and cellular membrane antigens. They are often used to select "lead compounds" in clinical trials. They help establish optimal doses of drugs used in these trials and provide objective evidence of the response to treatment. Molecular imaging of different regions of the body is needed because most diseases do not manifest abnormalities in circulating blood. Millions of biochemical processes go on continuously throughout the body that are not reflected in the circulating blood. Therefore, only a small percentage is detected by blood tests or other tests of body fluids.

The U.S. pharmaceutical industry spends nearly $50 billion a year in research and development of new drugs. Costs are relatively stable in the preclinical phase, but they have risen dramatically in clinical testing both in direct costs incurred and in time required to carry out the trials. Thirty-eight percent of compounds for which an Investigational New Drug Application (IND) is filed with the Food and Drug Administration (FDA) fail at the phase I stage. Sixty-three percent of those that enter phase II trials fail, and 45% of those that make it to phase III trials fail. Twenty-three percent of INDs are never approved nor reach the market.

For every 1,000 INDs filed, 620 compounds pass phase I. Of these, 229 pass phase II, and 126 pass phase III. Only 97 make it through the entire review process and are approved. Failure in phase III is a particularly disastrous result economically. Phase III is the most expensive point at which a drug can fail.

The number of drugs in phase III trials has remained at 375–400 over the past decade, despite significant increases in the number of drugs in preclinical, phase I and phase II trials. Some PET and SPECT tracer studies provide surrogate markers for assessing new drugs for treatment. This can reduce the cost of drug design and development. Today, approval of new drugs by the FDA requires clinical trials that show a statistically significant effect on mortality, which requires expensive, long-term studies. One of the problems is weeding out the "losers" from new drugs along the path of clinical trials. Pfizer, e.g., spent $1 billion to get Torcetrapib, an anticholesterol drug, to late-stage testing only to discover dangerous side effects that forced it to abandon this potential blockbuster. Molecular imaging can play a major role throughout drug design and discovery of new, effective drugs. They make clinical trials shorter and less expensive.

Diagnostic radiotracer studies are used initially to characterize the patient ("make the diagnosis") and help select homogeneous groups of patients for clinical trials of stable new drugs. Pretargeting nonradioactive molecules can be administered before the diagnostic tracer dose to increase the accumulation of the subsequently administered therapeutic drug. FDA regulations require approval of the pretargeting dose as well as the subsequent radioactive tracer drug.

People have taken drugs that effect mental activity since prehistoric times, most often alcohol and opiates. One third of all prescription drugs in the United States are chemicals given to affect mental activity which can be made more effective by individualization of treatment by molecular imaging, or the use of dedicated simple, single detectors.

According to economist Paul Romer, increasing knowledge is the way to increase economic returns. There are many more patients who could benefit from drugs developed from molecular imaging studies. The number of molecular imaging studies in every hospital is likely to increase. A very large number of tracer doses can be produced by a hospital cyclotron, which will decrease the cost of individual doses.

Genomics and molecular imaging make possible better selection of patients for "personalized" drug therapy. Patients diagnosed as having Alzheimer's disease often have other molecular manifestations that require specific treatment. Molecular imaging can greatly reduce patient care, and can easily justify itself economically. A minimum number of studies must be performed to cover the fixed costs of establishing and operating the tracer production facility and operate the PET scanners. Increasing the number of studies per day reduces the cost of each study.

Today, many potentially useful radiopharmaceuticals are still unapproved by the FDA. Regulatory approval is complex, expensive and takes a long time for both radioactive and stable drugs. Only 15 new stable pharmaceuticals received approval for sale in the United States in 2002. The peak number of approved drugs was 53 in 1996. The high cost and long time have increased recently, and the number of candidate drugs is now quite small. Innovation has been impaired, and many candidate compounds have failed to progress to approved drugs. Investors in pharmaceutical development are losing interest, and the pipeline for candidate drugs is distressingly empty.

Large pharmaceutical companies focus on developing "blockbuster" drugs with sales of billions of dollars. They are not interested in developing radiopharmaceuticals for small numbers of patients in whom molecular imaging makes possible personalized, custom-tailored therapy. They believe, erroneously, that the market is too small to justify the approval costs. This is true today, but will not be in the future.

The FDA has an accelerated approval program that makes new drugs available to patients with serious or life-threatening disease early in the regulatory process. Favorable response to the drug can be documented sooner than would be possible with a long clinical trial. Molecular imaging can be used, together with other tests, such as blood cell counts and chromosome evaluation, rather than rely on documentation of a favorable effect on long-term survival.

A pharmaceutical company was required to follow patients for 2 years to confirm long-term benefit on the survival of patients with leukemia who had shown a reduction in the percent of abnormal chromosomes in bone marrow and decreases in abnormal circulating white blood cells. The drug was given full approval after the long-term survival was found to be 91%.

An example of a drug approved under the accelerated approval program of the FDA is Gleevec. In May 2001, it was approved for treatment of chronic myeloid leukemia (CML), a life-threatening cancer that affects 40,000 people in the United States. Mark B. McClellan, M.D., Ph.D., at that time Commissioner of the Food and Drug Administration, said, "Our experience with Gleevec demonstrates the value of making promising drugs available early to patients with life threatening diseases based upon valid surrogate endpoints, such as short term tumor response rates, that are reasonably likely to predict that the drug can improve their lives … Our experience demonstrates the importance of follow-up studies after approval to confirm the drug provides a clinical benefit." This approach can be used to obtain approval of new radiopharmaceuticals.

Accelerated approval was given for Iressa, a drug that blocks the enzyme, tyrosine kinase, and decreases the growth of cancer cells. The drug blocks receptors of epidermal growth factor on the surface of cancer cells. Overexpression of the HER2 receptor occurs in aggressive forms of cancer of the bowel, breast, lung, head and neck, ovary, prostate, sarcoma, bladder, and kidney.

Glaxo Smith Kline, Inc., has >1,000 different compounds and 48 programs in their development pipeline. Their goal is to get past the early stages of development as rapidly, efficiently, and inexpensively as possible. One third fail for lack of efficacy. One third are found to be toxic. One third fail for regulatory, pharmacokinetic, or safety problems. What is needed is a better, faster, and cheaper way to move potentially useful radiopharmaceuticals forward to become approved drugs. Discoveries in academic institutions should not be the end but the beginning.

12.1. "Probe" Detectors

One does not always need expensive imaging devices to answer questions and solve problems. Simple "probe" detector devices can look at the entire brain or be focussed on specific regions, such as the frontal lobes or cerebellum. Probes for use with positron-emitting tracers (Fig.12.1) are 20–30 times less expensive than PET scanners. Because of the sensitivity of the detectors, and because a high degree of spatial resolution is not needed, low doses of the tracer can be given, so that the radiation dose to the patient is reduced by orders of magnitude compared with high-resolution imaging. In the latter case, there are statistical limitations, related to the number of counts per picture element. A probe can increase greatly the number of patients who can be studied at a much lower cost.

In 1986, Bice, Wagner, Frost et al. developed a simplified detection system for neuroreceptor studies in the human brain (Bice et al., 1986). The device, called *HEADS*, was patented 1988 (patent 4,712, 561), but never licensed for commercial development. In 1953, Brownell and Sweet had used a single pair of simple radiation detectors for localization of brain tumors (Brownell and Sweet, 11, 40, 1953).

In February, 1994, we carried out a clinical trial sponsored by the Pharmaceutical Products Division of the company OHMEDA. The purpose was to compare two antagonists of μ opiate receptors, nalmefene and naltrexone, that blocked opiate receptors in the human brain. Both drugs had been proposed to wake up patients after surgery performed under narcotic anesthesia. The challenge was to avoid postoperative respiratory depression caused by the anesthetic Fentanyl, an agonist of opiate receptors. The half-life of action of naloxone was about 1 hour, which was too short and the patients had to be monitored continuously to be sure that they would not stop breathing as the naloxone wore off, leaving fentanyl still suppressing respiration. Nalmefene was thought to have a

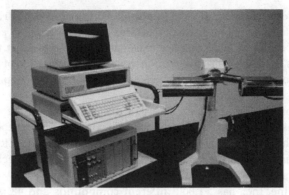

Fig. 12.1 Dual detectors ystemfor me asuringp ositron-emittingt racersi nt heb rain.

longer effective action than naloxone, but the FDA insisted that this be proved before they would approve the use of nalmefene.

The only way previously that one could assess the effect of drugs, such as naloxone, nalmefene, and other drugs on the brain, was to measure blood levels of the drug, or assess the subjective relief of pain. Because nalmefene blocked opiate receptors widely distributed in the brain, it was not necessary to use expensive imaging devices such as PET. A simple dual probe system could be used and was much less expensive, and much simpler.

The great sensitivity of the dual probe system made possible the administration of lower dose of the tracer. It was possible to carry out 20 studies at different times in a single person with the same radiation dose associated with a single PET scan. This reduced cost, improved statistical validity, and was far simpler to apply.

The study of a group of normal volunteers indicated that nalmefene blocked opiate receptors as measured by C-11 carfentanil binding, with a half-time of 24.3 hours, whereas naloxone manifest a shorter period of blockade, with a half-time of 9.0 hours. When the data were submitted to the FDA, the claim in the package insert that nalmefene had a longer duration of action was accepted. Unfortunately, the simple dual-detector approach never achieved widespread use, primarily because of the wide acceptance of PET and SPECT. Psychiatrists and other professionals will in the future use simple devices, analogous to the use of a blood pressure cuff, to examine the chemistry of patients' brains when a high degree of spatial resolution is not needed. A high-resolution imaging system may often not be needed to measure the quantitative effect of drugs that are antagonists of opiate receptors.

One can use a low spatial resolution detector system, because the naltrexone blocks all μ opiate receptors in the brain regardless of their location, and they are widespread throughout the brain. Probe devices for use with positron-emitting tracers have been used frequently to study mice, rats, dogs, baboons, and humans. In living mice, it was possible to obtain a dose–response curve for haloperidol on the binding of $[^{11}C]N$-methyl spiperone. Similar dose/response curves were obtained in humans, and the results correlated well with PET imaging studies of the same persons.

The use of simple probes can be used to study many important drugs involved in neurotransmission. For example, one could use probes in drug rehabilitation programs. Serotonin has been linked to violent, aggressive activity. Drugs that increase serotonin in the brain have been found effective in controlling violent impulses by compensating for serotonin deficiencies.

The brain chemistry of "happiness" can now be a focus of biological research, just as it has become a part of economic theory, sociology, psychology, and psychiatry. Jeremy Bentham defined happiness in terms of "utility" i.e., the sum of good emotions minus the bad. Human beings strive by their actions to increase pleasure and avoid pain. We can measure and correlate objective measurements of brain chemical processes with the emotions of pleasure, fear, and pain, and perhaps modify the undesirable brain chemistry when it is not helpful. Simple, probe radiation detectors can be used in many of these studies.

Drugs that block binding sites include selective serotonin reuptake inhibitors (SSRIs). In 1970, the Eli Lilly Company developed fluoxetine (Prozac). Their research began with nizoxetine, a selective norepinephrine reuptake inhibitor. Of the compounds that inhibited reuptake of serotonin, norepinephrine, and dopamine, fluoxetine was the most potent and selective inhibitor of serotonin reuptake. It was approved for use by the FDA in December, 1987, and used to reduce excessive serotonin secretion from carcinoid tumors.

Monoamine oxidase inhibitors, such as SSRIs, have been shown to be effective in the treatment of depression, and they have become among the most widely used prescription drugs in the United States. Prozac is used not only to treat major depressive disorders but also bulimia nervosa, obsessive-compulsive disorder, panic disorder, and premenstrual dysphoric disorder. Multiple serotonin receptor subtypes are involved. Specific serotonin receptor subtype agonists and antagonists have been radiolabeled with positron-emitting tracers to assess the state of the serotonergic system.

According to the National Center for Health Statistics, prescriptions for antidepressants in the United States climbed to 40 per 100 physician office visits in 2004, from 13.8 in 1995. New Zealand and the United States are the only developed countries that allow direct-to-consumer advertising of brand-name prescription drugs. Pharmaceutical companies in the United States spent $4.2 billion for this advertising in 2005. Today, patients must take antidepressant drugs for 4 to 6 weeks before they know whether or not the medication will work.

Fred Ovsiew of the University of Chicago said, "We are only at the beginning of trying to develop treatments that are rationally based on [different patterns of brain abnormalities]." Molecular imaging helps reach this goal.

New radiopharmaceuticals are reviewed by Radioactive Drug Research Committees at universities or research laboratories. In 2003, there were 284 such research studies in the United States involving 2,797 human subjects and >120 different radioactive molecules. Efforts are being made to simplify toxicity studies. Radiotracers are often radiolabeled natural body constituents, administered in micromolar or millimolar quantities, far less than one hundredth of the toxic dose.

The triad of tools in drug design and development are genetics, pharmacology, and molecular imaging. Genetic profiling can identify persons at risk of developing disease, and subsequently detect abnormal molecular processes as they occur, and intervene before symptoms occur.

Drugs are generally of two types: those that stimulate a molecular process (agonists) or those that suppress it (antagonists). Peptide and proteins can be screened by high-throughput screening early in the process of drug design. Molecular manifestations have become the new language of medical diagnosis.

Molecular medicine can play an important role in the selection of patients for clinical trials. The criteria for selecting patients with Alzheimer's disease can be clearly defined. Current standards from a National Institutes of Health working group in 1984 are out of date. Using the current criteria, patients are selected as having probable Alzheimer's disease if they have trouble with their memory and impairment of at least one other mental function. To make the diagnosis of Alzheimer's disease "both of these problems must interfere with social function or the activities of daily living." The diagnosis requires that a person "suffer memory loss that gets worse over a six-month period" and then "at least one physical 'biomarker'" for confirmation. An MRI scan can now show shrinking of a particular part of the brain, or a PET scan can reveal decreased metabolic activity in characteristic parts of the brain, such as the temporoparietal lobes, or abnormal proteins in the cerebrospinal fluid, or a genetic mutation linked to the disease. Current research is directed toward imaging of deposits of amyloid protein in the brain, by using specific radiotracers.

Search continues for plasma biomarkers that could be used to predict which patients diagnosed as mild cognitive impairment will go on to develop Alzheimer's disease. Tony Wyss-Coray and colleagues at Stanford University have found 18 plasma markers that can correctly classify patients with Alzheimer's disease compared with patients with other types of dementia or controls.

Since the late 1980s, more than a dozen major pharmaceuticals companies, including Upjohn, Bristol-Myers Squibb Pharmaceuticals, and Merck Sharp & Dohme, have joined the multibillion-dollar race to find drugs capable of blocking the production of beta-amyloid. Before a cell can release β-amyloid, it must be cut loose from the rest of the Alzheimer's amyloid precursor (AAP) protein by an enzyme. Blocking the action of this enzyme is the main goal of researchers at Athena Neurosciences, which, with backing from the drug company giant Eli Lilly, has built up one of the world's largest programs of research on Alzheimer's disease. Ivan Lieberburg, who heads this program, says that his scientists have identified the enzyme and, more importantly, have found several compounds which block its activity in the test tube.

Despite such apparent successes, however, the β-amyloid hypothesis has recently experienced some setbacks. The first stems from efforts to make a mouse model of Alzheimer's disease, which would be a boon to drugs testers. In 1991, three teams succeeded in genetically engineering mice to overproduce APP in their brain cells. One of these teams claimed that their transgenic mice developed brain lesions similar to those found in Alzheimer's patients. For many, this was proof, if any were needed, that APP causes Alzheimer's disease by unleashing β-amyloid on the brain. It seemed that all that remained was to find out how β-amyloid kills brain cells.

But in March of last year, the paper was retracted from *Nature*, after its authors were unable to replicate their earlier findings. What's more, there were calls for a fraud investigation, after allegations that photographs purporting to show brain damage in transgenic mice were actually photographs of brains autopsied from Alzheimer's patients.

Within months came another challenge to the β-amyloid hypothesis. Up until last year, most researchers had assumed that β-amyloid is only produced when something goes wrong with the way APP is processed in cells. But in September and October came evidence to the contrary. Three research teams, led by scientists at Harvard University, Athena Neurosciences, and Case Western Reserve University in Ohio, discovered that a soluble form of the β-amyloid peptide–that is, a form which does not aggregate into clumps of fibrils—could be detected in the spinal fluids and bloodstreams of perfectly healthy people. This by no means demolished the idea that β-amyloid is involved in Alzheimer's disease, because the disease could still be caused by unusually high levels of β-amyloid or faulty processing of the peptide by cells. But it did complicate the picture significantly.

The most important controversy, however, has centered on β-amyloid's guilt (or innocence) as a neurotoxin. In the summer of 1990, the issue looked settled. Based on experiments with a specific fragment of β-amyloid, a team led by Bruce Yankner at Harvard University claimed that β-amyloid could indeed kill neurons provided the neurons had stopped dividing. But subsequently, several respected laboratories reported difficulties replicating this result, and for a time the debate became heated. Now the tide seems to be turning in Yankner's favor. The key, argues Dennis Selkoe, who runs another Alzheimer's laboratory at Harvard, is that β-amyloid only becomes toxic after it has aggregated into insoluble fibrils. "People have had real problems reproducing in vitro toxicity by just adding beta-peptide in solution to media," he admits. "But if they allow the peptide to aggregate, it looks like it does produce some local toxicity when added to cultured cells."

McGeer and Rogers are willing to concede that β-amyloid can kill neurons directly, but they think that most of the damage is caused by the peptide's immunological effects, specifically, its ability to activate the complement cascade. Their unpublished results show that "when immune factors are present, the toxicity of beta-amyloid jumps tremendously," says Rogers.

Proponents of the β-amyloid hypothesis tend to believe the reverse is true–that β-amyloid itself does most of the cell killing. But most acknowledge that there could still be a role for anti-inflammatory drugs in slowing the progress of Alzheimer's disease. "I think that has to be considered as a possibility," says Yankner. "So I'm on the fence about that right now."

Lieberburg says his company has a foot in both camps. "There is good reason to suspect that inflammation is part and parcel of Alzheimer's disease, and contributes to the ongoing dementia that the patient experiences." He notes that the side effects of indomethacin will probably make it unsuitable for widespread use in elderly patients, but he concedes that the results from the clinical trial are very interesting and congratulates McGeer and Rogers for forcing open a new avenue of Alzheimer's research.

"It's people like Joe Rogers and Pat McGeer who continued to hammer away on this inflammation issue and to say, 'Look, this is right under your noses here; this looks like chronic arthritis of the brain–why do you continue to ignore this?'" In fact, confides Lieberburg, Athena has been pursuing its own secret anti-inflammatory program for the past 4 years, and has been screening for a drug, which can penetrate the blood-brain barrier while having better anti-inflammatory properties and fewer side effects than indomethacin.

McGeer and Rogers have obtained a U.S. patent on the use of drugs such as indomethacin to treat Alzheimer's disease. Pharmaceuticals companies, although expressing interest in their work–"my calendar this spring has been filled with visits to drug companies," says Rogers–have not yet been willing to help the two researchers through the long and expensive process of winning approval for indomethacin as an Alzheimer's treatment from the FDA. But the two are forging ahead anyway. Rogers, with financial support from local doctors in Sun City and the Sun Health Research Institute, has begun a new clinical trial of indo-methacin, this time in 120 Alzheimer's patients.

The legacy of McGeer and Rogers' research might extend well beyond Alzheimer's disease. McGeer believes that the way in which β-amyloid provokes the immune system, to such fatal effect, may hold an important lesson for researchers investigating other autoimmune diseases. According to the conventional view, autoimmunity occurs when the immune system mistakenly reacts to specific proteins, or antigens, as though they were foreign. As a result, the immune system produces antibodies and immune cells that are directed against some of the body's own antigens.

In the case of Alzheimer's disease, however, and perhaps other autoimmune conditions, McGeer envisions something different happening. Instead of stimulating the production of antibodies and immune cells that home in on specific antigens, β-amyloid clumps seem to activate the immune system at a deeper level. Antibodies normally stimulate the complement cascade, but β-amyloid seems to "short-circuit" this process, stimulating the cascade without the need for antibodies. McGeer argues that just because one or two autoimmune diseases–such as myasthenia gravis, a muscle weakening disease–are known to be triggered by a specific antigen, we should not assume that the rest work the same way. "No one has been able, for example, to define antigens in rheumatoid arthritis, Behcet's disease, temporal arteritis," he says. "I think there's a missing link here, and quite a big one."

13

SubstanceAbuse

On October 19, 2007, a freshman at the Westminster Choir College in Princeton, NJ, a talented cellist and pianist, injected himself with heroin provided by a classmate. He passed out and was driven to a friend's apartment where he lay unconscious for several hours. His friends noticed that he was not breathing and called 911. He was found dead when help arrived.

Throughout history, people have known that extracts of certain plants could make them "feel good." Positron emission tomography (PET) studies with F-18 deoxyglucose (FDG) have shown that all drugs that induce feelings of well being reduce neuronal activity in nearly all regions of the brain, perhaps suggesting that our brains are in a nearly constant state of alertness.

Nitrous oxide was produced in 1772 by the English chemist and Presbyterian minister Joseph Priestley, and it was found to be mind altering. Many people inhaled what was called "laughing gas" because it produced "a highly pleasurable thrilling." For nearly a century, nitrous oxide was the drug of choice for young people. Theaters put on "laughing gas evenings," where theatergoers could inhale the gas and amuse their friends as they staggered around. Parties were held that featured inhaling the gas. Not until the middle of the nineteenth century was it discovered that inhaling nitrous oxide could relieve pain. Its use as an anesthetic was later replaced by ether and chloroform. Nitrous oxide was the first example of the balancing of pleasure against pain, which some believe is the essence of life.

For thousands of years, resins from poppy plants were ingested to produce a pleasurable effect. In paintings, Hypnos, the Greek god of sleep, is shown carrying poppy plants. The Roman god of sleep, Somnus, is pictured carrying the plant. The Greek physician, Galen, wrote that opium "cures headaches, dizziness, deafness, epilepsy, poor vision, asthma … tightness of breath, colic, jaundice, gall stones, urinary disorders, fevers, dropsy, leprosy, the troubles to which women are subject, and melancholy."

Arab, Greek, and Roman physicians used opium before the sixteenth century, when European medicine officially adopted its use, and it became its most widely used medication. Thomas Sydenham, England's most famous physician, improved Paracelsus' synthesis of *laudanum*, an alcoholic extract from poppies. Sydenham wrote, "I cannot forebear mentioning with gratitude the goodness of the Supreme Being, who has supplied afflicted mankind with opiates for their relief [of pain], no other remedy being equally powerful to overcome a great number of diseases, or to eradicate them effectually."

The abuse of opiates has a long history. Thomas De Quincy, in his book, *Confessions of an English Opium Eater*, describes how he was able to buy laudanum without a prescription in London in 1804. Laudanum is a tincture of opium in alcohol. A painting by Andre Gide entitled *The Pleasures of Constantinople* depicts an ancient Turk smoking opium.

Cocaine, like opium, also was widely dispensed at pharmacies in solutions of alcohol. Queen Victoria was fond of marijuana. Thomas Jefferson smoked marijuana grown on his plantation. He believed that marijuana should be sold at grocery stores the same as beer or wine.

Billions of dollars are invested every year to develop drugs that affect behavior. A third of all the prescription drugs taken in the United States are given to affect mental activity. Millions of people ingest narcotics, tranquillizers, stimulants and sleeping pills, often harming themselves.

Morphine, heroin, and methadone bind to opiate receptors to varying degrees, reflecting their chemical affinity.

Methadone is a synthetic opioid, developed in Germany in 1937 and first marketed in the 1980s as an analgesic and later for the treatment of narcotic addiction. Methadone has been the nation's mainstay addiction therapy since the 1970s. It is dispensed by about 1,100 clinics to roughly 250,000 people nationwide. To stem abuses, methadone clinics initially require addicts to show up daily for their doses.

Pentazocin was the first clinically approved drug that was both an agonist and antagonist of opiate receptors that would relieve pain. The antagonist action was to prevent addiction. It has been used to treat heroin addicts.

Buprenorphine, administered in methadone clinics, has both agonist and antagonist effects. In Florida, buprenorphine caused 312 deaths in the first six months of 2006, more than any other drug.

In 1999, methadone was the cause in 13% of all drug overdose deaths in the United States. The number of deaths from methadone in the United States in 2004 was 3,849, a 390% rise since 1999.

Cocaine kills more people than methadone. Most patients who have died from methadone overdoses were taking the drug for pain, not addiction, and often took it with other drugs.

Methamophetamine (meth) came on the scene in the early 1970s, and home production is becoming more and more widespread in Europe, especially in the Czech Republic, where there are 416 production operations. According to the United Nations, the Czech Republic has extremely high levels of cannabis use, and the highest for the drug Ecstasy. By far the Czech's have the worst methamphamine abuse in Europe. Meth can be produced from pseudoephedrine found in common cold medications. It is half the price of cocaine. In the United States, significant success has been achieved in the fight against meth producers, but Asian and Mexican laboratories continue to supply American users.

Cocaine and heroin are derivatives of plants, unlike other drugs, such as Prozac, Xanax, and barbiturates. Richard Degrandpre (*The Cult of Pharmacology*, Duke University Press, 2007) has written that there are the good drugs (medicines) and bad drugs, sold on the street. The most dangerous drugs are legal medicines that are more toxic than many illegal drugs, including marijuana and Ecstasy.

Nearly half of county law enforcement officials consider methamphetamine their primary drug problem, more than cocaine, marijuana and heroin combined, according to a survey by the National Association of Counties. London and colleagues reported corticolimbic abnormalities in methamphetamine-dependent subjects during early drug abstinence (London et al., 2004).

Opiates remain the most effective method for the relief of pain. Witold Gombrowicz has written, "No matter what we are told, there exists in the entire compass of the Universe, throughout the whole space of Being, one and only one awful, impossible, unacceptable element, one and only one thing that is truly and absolutely devastating: pain. It is on pain and on nothing else that the entire dynamic of existence depends" (*New York Review of Books*, January 12, 2006).

Thomas DeQuincey (1785–1859) in his book *Confessions of an English Opium-Eater* stated his belief that laudanum, when taken to ease the pain of his toothache was "the secret of happiness." He wrote, "My vanished pain is now a trifle in my eyes ... Here was the secret of happiness, that which philosophers had disputed for so many ages ... happiness that might now be bought for a penny, and carried in the waist-coat pocket; portable ecstasies that might be corked up in a pint bottle; and peace of mind that could be sent by the mail."

In the early 1970s, Hans Kosterlitz and John Hughes, two Scottish scientists, discovered opiate receptors in the brain that bound opiate-like substances produced by the brain itself— the brain's own morphine. Using newly developed high performance liquid chromatography, they discovered peptides isolated from the brain, the pain-relieving effects of which could be blocked by the opiate receptor-blocking drug, naloxone. Avram Goldstein, Professor of Pharmacology at Stanford University, discovered one of the three families of opioid peptides—the dynorphins, which Eric Simon called endorphins. Hughes and Kosterlitz preferred the name enkephalins, from *en kephalos* (in the head). There are different endogenous opioid peptides produced in the brain, with different receptor specificities, including μ, δ, ϵ, and κ receptors. They are involved in perception of pain, the immune response and other processes.

Taking drugs to affect mental processes raises moral, ethical, and legal, as well as biomedical issues. Everyone wants to increase pleasure and avoid pain. Drugs that relieve pain increase pleasure in the absence of pain. Criminal activity often involves pain/ avoiding, pleasure-seeking drugs. Early in this century, the U.S. government tried to eliminate the illegal use of morphine, heroin, and other opiates, but they failed. Governmental prohibition aggravated the situation. The cure for opiate addiction turned out to be far worse than the disease, just as the prohibition of alcohol in the United States had similar, devastating consequences.

Abuse of prescription drugs is as big a problem as the abuse of illicit narcotics. Painkillers, sedatives, and other medicines are overprescribed, according to the United Nations affiliated International Narcotics Control Board in 2006. Up to 50% of all drugs taken in developing countries are counterfeit. Analgesics are the main injected drugs of abuse. In tablet form in France, 20–25% of the commercial drug (Subutex, Suboxone) is diverted to the black market.

The hexagonal orange pills buprenorphine users call "bupe" were introduced to treat heroin and pain-pill abuse. (Fred Schulte and Doug Donovan, *Baltimore Sun*, December 16, 2007). Making buprenorphine widely available made it easy for patients to sell the narcotic illegally, leading to growing abuse. Addicts use buprenorphine to tide them over when they cannot obtain heroin or other narcotics.

Suboxone was introduced for treating addiction as a chronic health condition, similar to treating diabetes with insulin. The pills relieve addicts' cravings for opiates. Illegal sales and abuse remain far less than other abused narcotics but are on the rise. Rolley E. Johnson, Vice President for Scientific and Regulatory Affairs for the company Reckitt Benckiser, said that abuse is inevitable. "Anything that has opioid-like effects, which buprenorphine does, can and will be abused by those people seeking that effect," said Johnson, himself a buprenorphine researcher.

The number of Americans abusing prescription drugs doubled from 7.8 million in 1992 to 15.1 million in 2003. The painkiller hydrocodone was used by 7.4% of college students in the United States in 2005. It is a semisynthetic opioid derived from two of naturally occurring opiates, codeine and thebaine. Production of this drug has increased in recent years. In Scandinavia, flunitrazapam, a sedative, is sold as Rohypnol, and it is widely known as a "date-rape drug."

Those who advocate decriminalization of taking illicit drugs believe that virtually overnight there would be a plummeting of the price of the drugs. Street crime, money laundering, gang violence, and corruption in law enforcement that involves drugs, would disappear. Organized crime and drug cartels would fall. Street dealers would disappear. There would be fewer long-term prison sentences, and fewer nonviolent offenders put in jail. The breakdown of families and communities would decrease. A temporary rise in drug abuse would be expected to level off as a result of addiction–prevention programs. Billions of dollars used for drug enforcement could then be used for treatment.

Suboxone is the centerpiece of a government effort to shift opiate addiction treatment away from restrictive clinics to private doctors' offices. The federal government spent at least $26 million to help develop buprenorphine in partnership with Reckitt Benckiser Pharmaceuticals Inc., a subsidiary of a British company. The pills reduce craving for opiates and ease the sickness that addicts feel when they stop "using." It makes them feel "normal."

13.1. Imaging Opiate Receptors

On May 23, 1984, after parking my car in the parking lot on the east side of the School of Public Health at Johns Hopkins, I crossed Wolfe Street to enter the hospital. I took an elevator to the Anesthesiology Department, where I was to be injected with carfentanil, a drug that had been given to wild animals, such as black bears and bison, to immobilize them, but never before to humans. The goal of our research was to obtain a better understanding of the role that the opioid system played in pain, emotions, and addiction.

Before I was injected with carfentanil, we had carried out studies in rodents and baboons. The dose that I was to be given would be exceedingly small, but we thought it wise to have an anesthetist and an anesthesia machine present in case my breathing stopped. As it turned out, there was no effect on my breathing, but there was constriction of my pupils. We scheduled a second injection and PET imaging procedure for the next day in the nuclear medicine department.

As I lay with my head in the NeuroEcat PET scanner, I was injected with 25 mc of C-11 carfenanil, the label on 80 ng kg^{-1} of the drug. In the room were Jim Frost, Sol Snyder, Jon Links, our chief nuclear medicine technologist, Jim Langan, who had been in nuclear medicine at Johns Hopkins since its inception in 1958, and two other technologists, Jay Rhine and Martin Stumpf.

Carfentanil is the most potent of all opioid drugs, and it is potentially toxic or even lethal to humans. People have developed symptoms of drug toxicity from consuming animals given immobilizing drugs. As little as 20 µg (an almost invisible drop) of carfentanil is lethal to humans. It is not wise to eat animals that have been immobilized with carfentanil

The PET studies with carbon-11 carfentanil derived from the study in 1973 by Pert, Snyder, and colleagues at Johns Hopkins who used autoradiography with C-14-labeled drugs to localize opiate receptors in the rat brain (*Science*, March 9, 1973). Two other research groups—one group led by Dr. Eric J. Simon at New York University and the other by Dr. Lars Terenius at Uppsala University in Sweden—reported within a short time that brain cells had membrane receptors that seemed tailor-made to latch onto opiates like morphine. The opiates exert their effect on the brain by binding to those receptors.

In the early 1970s, Hans Kosterlitz and John Hughes, two Scottish scientists, discovered opiate receptors in brains that bound opiate-like substances produced by the brain itself—the brain's own morphine. Using newly developed high-performance liquid chromatography, they discovered peptides isolated from the brain, the pain-relieving effects of which could be blocked by the opiate receptor-blocking drug naloxone.

These investigators found that receptors in some smooth muscle in the guinea pig intestine and in the mouse vas deferens, the duct that carries sperm, have opiate receptors, and that the potency of opiates or opiate-like chemicals can be measured by how well they inhibit muscle contraction.

The endogenous opiates, called endorphins, affect the brain in many ways, sometimes acting as the chemical messengers between neurons and sometimes modulating the messages carried by other messengers. They affect mood and hunger as well as pain perception.

Solomon Snyder, director of the department of neuroscience at Johns Hopkins, has said, "Not only did Dr. Kosterlitz discover the endorphins, but he found that there are subtypes of opiate receptors. That's one of the biggest breakthroughs in the field. If there are different kinds of receptors, one kind might be for pain relief while another causes addiction. That could help the pharmaceutical industry design drugs to do the good without the bad, making it possible to try to develop nonaddicting opiates."

It was the work of Pedro Cuatrecasas and Jesse Roth at the National Institutes of Health (NIH) in Bethesda, MD, who in 1970, characterized insulin receptors, and inspired Solomon Snyder to search for opiate receptors using naloxone, known to block the action of opiates.

Naloxone blocks receptor binding of endogenous opioids, called endorphins or enkephalins, produced by the pituitary, as well as blocking the effects of administered narcotics, such as morphine. These blocking drugs are used to awaken patients operated on under opiate anesthesia. Persons in coma from ingesting illegal narcotics, such as morphine or heroin, wake up within seconds after administration of naloxone.

Frost and colleagues showed that [^{11}C]diprenorphine also could be used to image opiate receptors in human beings. Sprenger and colleagues at the Technical University of Munich (*Eur. J. Pain* 9:117, 2005) observed that the thalamus, prefrontal and cingulate cortex, basal ganglia, and midbrain structures possess large numbers of opiate receptors.

In the first study with C-11 carfentanil, three planes through my brain were imaged: the cephalic plane passed through the cerebral cortex above the corpus callosum; the central plane passed through the caudate nucleus, putamen, thalamus, and cerebral cortex; and the lower plane passed through the cerebellum. PET images were obtained an hour after injection of the tracer. After 30 minutes, the greatest accumulation of radioactivity was seen in the medial thalamus with less in the lateral thalamus. The frontal and temporoparietal cerebral cortex had less activity but substantially more than the occipital cortex. In the caudate slice, activity was seen in the pituitary and temporal lobes of the cerebral cortex. Three hours later, naloxone was injected at a dose of 1 mg kg^{-1}, followed by another dose of 25 mc of C-11 carfentanil. The PET study showed that the radioactivity accumulation was markedly decreased compared with that in the earlier study before naloxone. This provided strong evidence that the carfenanil was binding to μ opiate receptors.

An article was printed in the *Baltimore Sun* on Tuesday, September 26, 1983: "Johns Hopkins medical scientists have opened a new window into the living brain to study the causes of some of the most disturbing mental and physical disorders of human beings … Previously, the mapping of these tiny, but immensely important brain parts had been limited to tissues obtained at autopsies … A lot of changes can occur at or after death, so the results of postmortem studies aren't always reliable."

Kenneth Casey carried out PET studies of the brain with F-18 FDG during stimulation with heat (40 or 50°C) applied to each subject's forearm, increasing the degree and duration of heat until it produced pain (*J. Neurophysiol.* 85: 951–959, 2001). Glucose use increased in cortical regions, the thalamus, the vermis, and the paravermis of the cerebellum. During repetitive noxious heat stimulation, cortical activation occurred before activation of subcortical regions.

Lorenz and others at the Institute of Neurophysiology and Pathophysiology in Hamburg-Eppendorf, Germany, found that pain activated the limbic forebrain, including the insula, perigenual anterior cingulated ventral striatum, and prefrontal cortex (Mind/Brain Symposium, 2004).

The pain of chronic rheumatoid arthritis, trigeminal neuralgia and central poststroke pain is associated with decreased [^{11}C]carfentanil binding whenever the patients have pain, but not during the pain-free intervals, suggesting the release of endogenous opioids in response to pain. There also may be internalization/down-regulation or loss of neurons carrying opiate receptors.

PET imaging of opiate receptors can determine which of the four different opiate receptor subtypes a drug affects. δ Opioid receptors are located in the limbic system and pons. κ Opioid receptors are in the nucleus accumbens and other corticolimbic areas. The limbic system and hypothalmus, as well as the cerebral cortex, seem to be involved in the emotions of violence, anger, and hatred, in addition to their role in thinking and reasoning. Many neuronal systems connect the primarily emotional and cognitive regions of the brain. No single region of the brain is involved in the initiation of fear, anger, greed, arrogance, and aggression, but there is increasing evidence that there is a relationship between monoamine neurotransmitter—epinephrine, serotonin (5-hydroxytryptamine [5-HT]), and dopamine—and anger, aggression, greed, arrogance, and violence.

The emotions of violence, aggression, greed, arrogance, and rage are related to the emotion of pain. Autonomic functions, such as heart rate, hormone release, and other physiological activity related to emotions are controlled by the hypothalamus. Epinephrine is involved in excitement, fear, and anger. When we encounter something unfamiliar, intrusive, or threatening, the hormones, epinephrine, and norepinephrine are secreted, affecting the amygdala, the limbic system, and the prefrontal cortex. 5-HT is involved in ㅁㅁㅁㅁㅁㅁㅁㅁㅁㅁ ㅁㅁㅁㅁㅁㅁㅁㅁ ㅁㅁㅁㅁㅁㅁ ㅁㅁㅁ ㅁㅁㅁㅁ ㅁㅁㅁㅁㅁ ㅁㅁ ㅁㅁㅁ ㅁㅁㅁㅁ ㅁㅁㅁㅁㅁㅁ ㅁㅁㅁ *Psychiatry* 46:35–44, 2001).

The insula brings information related to emotions from the autonomic nervous system to consciousness. It lies between the temporal and parietal cortex, and it plays a major role in addiction. It receives information from the thalamus and sends it to other limbic-related structures, such as the amygdala, the ventral striatum, and the orbitofrontal cortex. The anterior insula is related to olfactory, gustatory, vicero-autonomic, and limbic function, whereas the posterior insula is related to auditory-somesthetic-skeletomotor function. Functional magnetic resonance imaging (fMRI) indicates that the insula also plays a role in the experiencing of pain, as well as the emotions that include anger, fear, disgust, happiness, and sadness. fMRI studies indicate the insula in conscious desires, such as the craving of food or drugs. In essence, the insula is a major region of the brain that integrates information relating emotionally related autonomic functions to cognitive processes.

The endogenous endorphin system is involved in cocaine addiction as well. Cocaine addiction causes an increase in the number of the brain's opiate receptors, thought to be a response to a decrease in release of endogenous endorphins.

Chronic cocaine abuse causes the brain to reduce its production of endorphins and enkephalins. The more severe the craving for cocaine, the greater number of C-11 carfentanil binding receptors are seen in the frontal cortex, anterior cingulate cortex, caudate, and thalamus. To compensate for the drop in enkephalin production, the brain produces more opiate receptors to bind as much of the remaining enkephalin molecules as possible. This increase in the number of "empty" opiate receptors may be related to the feeling of craving for cocaine.

The narcotic, fentanyl, which is widely used as an anesthetic, has become abused. In Philadelphia, there were 53 fatal doses between April and June 2006. In Chicago, there were 100 deaths in 2006. Across the entire United States, there were at least 500 deaths in 2006. Fentanyl is often mixed with heroin to boost the high, and it is sold under the street names of Drop Dead, Lethal Injection, and Get High or Die Trying. In Baltimore, the heroin-addicted population is estimated to be 60,000. Fentanyl is sold on the street under the names China White and Tango & Cash.

Cocaine and amphetamine inhibit reuptake of dopamine, Amphetamine also causes release of presynaptic dopamine. Both drugs affect the nucleus accumbens. Opiates and tetrahydrocannabinol, the active principle in cannabis, bind to μ opioid and cannabin (CB) 1 receptors. The CB1 receptor is found chiefly in the brain but also in other regions. The CB2 receptor is found only outside the brain.

In the mesolimbic pathway, these receptors are located on GABAergic neurons in the ventral tegmental area (VTA). Activation of cannabin receptors causes disinhibition of the dopaminergic projection neurons, and dopamine is released in neurons along mesolimbic pathways. Alcohol also directly increases the firing rate of dopaminergic VTA neurons.

Using fMRI, Daniel Levatin found that listening to music brings about a cascade of brain chemical activity. Music activates the forebrain, which is involved in analyzing the structure of the music. Then, the nucleus accumbens and VTA increase dopamine production, which triggers a sense of well being. According to Levatin, the cerebellum is activated if the tune results in tension, brought about by sudden changes in melody or tempo.

There are three dopaminergic pathways in the brain: the nigrostriatal, the mesolimbic-mesocortical and the tuberculoinfundibular. In addition to being a neurotransmitter, dopamine is a precursor of noradrenaline and adrenaline.

Alcohol, psychostimulants (e.g., cocaine and methamphetamine), nicotine, and opiates, increase dopamine levels in the brain. The larger amount of dopamine give rises to feelings of pleasure, and is responsible for craving and drug dependence.

Withdrawal of the abused substance results in a decrease in the large quantities of dopamine resulting from the drugs. The substance-dependent person experiences symptoms of depression, fatigue, and withdrawal. The only way to relieve these symptoms is to use increasing quantities of the abused substance of abuse, reinforcing dependence on the drug.

Marijuana is the most widely used illegal drug in the United States. Many began smoking marijuana in the mid-1960s and early 1970s, but fewer start now.

Abused drugs produce changes in regional neuronal activity, indicated by studies of regional cerebral glucose and blood flow. Ronald Herning of National Institute of Drug Abuse Intramural Research Program measured blood flow through the anterior and middle cerebral arteries in 54 marijuana users (11 light, 23 moderate, and 20 heavy) and compared the results with observations of blood flow in 18 similarly aged adults who did not use the drug. The cerebral blood flow was significantly higher in all three groups of marijuana users than in the control group.

Wilson and colleagues from Duke University studied 57 subjects who had started using marijuana before age 17, and found that they had significantly higher cerebral blood flow than other males. These changes persist in heavy users of the drug even after a month of abstinence.

Volkow and colleagues from Brookhaven National Laboratory (BNL) found that marijuana abusers had lower overall cerebellar glucose metabolism than normal subjects, but they had increased glucose utilization in orbitofrontal cortex, prefrontal cortex, and basal ganglia.

Alcohol, like all addictive drugs, increases the brain's production of dopamine, which sends a strong reinforcing signal of pleasure and reward. Peter Thanos and colleagues from BNL reported that alcohol stimulates the pleasurable effects of marijuana (*Behav. Brain Res.*, September 2, 2005).

The brain responds to constant dopamine stimulation by "down-regulating," or decreasing, the number of D2 dopamine receptors. Alcoholics then drink more and more to try to override this blunted pleasure response. And people with initially low levels of these dopamine receptors are particularly prone to abuse alcohol or other drugs.

Thanos administered drugs to block the brain's cannabinoid receptors, known as CB1. These brain receptors are directly involved in triggering the reinforcing properties of marijuana, and are believed to play a role in stimulating reward pathways in response to alcohol.

Abused drugs, including alcohol, opiates, cannabinoids, and psychostimulants, such as nicotine, activate the mesolimbic dopaminergic brain reward systems and increase dopamine levels in the nucleus accumbens. Specific CB1 receptors modulate the dopamine-releasing effects of nicotine. Rimonabant (SR141716), a receptor antagonist, blocks the dopamine-releasing and rewarding effects of cannabinoids. Blocking CB1 receptors is ineffective in reducing self-administration of cocaine in rodents and primates, but it reduces the return of treated cocaine-seeking behavior.

ⵏⵓⵓⴷⵍⵓ ⵉ ⵓⵍⵓⵓⴰ ⵉⵍ ⴷⵓⵓⵖ, ⵗⴷⵓⵓⵓⴷⵍⵯⵍⴰⴰⵓⵍⵏⵓⵓⵍⵍⵓⵓ ⵢⵓⵓⴰⵓⵍⵓⵓⵓⵎⵯⵓⵍⵓⵓⵓⵎⵓⵓⵓ ⵟⵯⵓⵍⵯⵓⵍⵯⵉ ⵓⵟ ⵓⵓⵓⵟⴰⴰⵓⵒ ⵉⴰ a synthetic drug with psychedelic and stimulant effects. In 1988, it became a schedule I substance of abuse under the Controlled Substances Act. Trafficking in this drug can lead to 10 years in prison.

Cocaine and heroin are derivatives of plants, unlike other drugs, such as Prozac, Xanax, and barbiturates. Richard Degrandpre (*The Cult of Pharmacology*, Duke University Press, 2007) has written that there are the good drugs (called medicines) and bad drugs, sold on the street. The most dangerous drugs are legal medicines that are more toxic than many illegal drugs, including marijuana and Ecstasy.

Billions of dollars are invested every year to develop drugs that affect behavior. Pentazocin was the first clinically approved mixed agonist/antagonist that would relieve pain. For decades, this long-acting opiate agonist has also been given to treat heroin addicts.

Nearly half of county law enforcement officials consider methamphetamine their primary drug problem, more than cocaine, marijuana, and heroin combined, according to a survey by the National Association of Counties. London and colleagues reported corticolimbic abnormalities in methamphetamine-dependent subjects during early drug abstinence (London et al., 2004).

13.2. Ecstasy

According to the office of National Drug Control Policy, MDMA is predominantly used at all-night dances known as "raves," but its use is moving into private homes, high schools, college dorms, and shopping malls. MDMA is a stimulant with psychedelic effects. Side effects include confusion, depression, anxiety, sleeplessness, drug craving, and paranoia. Adverse physical effects include muscle tension, involuntary teeth clenching, nausea, blurred vision, feeling faint, tremors, rapid eye movement, and sweating or chills.

MDMA increases the levels of neurotransmitters such as 5-HT, dopamine, and norepinephrine by causing release of the neurotransmitters from their storage sites. It causes a significant depletion of neurotransmitters, including 5-HT. The number of new MDMA users has risen since 1993, when there were 168,000 new users. By 2001, this number had reached 1.8 million. The National Survey on Drug Use and Health found that 15.1% of 18- to 25-year-olds surveyed in 2002 had used MDMA at least once in their lifetime. There were 676,000 who had used the drug within the month before the survey.

The University of Michigan's Monitoring the Future Study found that MDMA use among high school students declined from 2002 to 2003. Among 10th and 12th graders surveyed in 1996, annual prevalence of MDMA use (use in the past year) was 4.6% in both grades. By 2003, annual prevalence had decreased to 3% among 10th graders and 4.5% among 12th graders In 2002, 12.7% of college students and 14.6% of young adults reported having used MDMA at least once. Approximately 6.8% of college students and 6.2% of young adults reported MDMA use in the previous year. Ecstasy is often taken with alcohol or marijuana, or both. It is also sometimes taken in combination or sequentially with various other legal and illegal drugs, including lysergic acid diethylamide (LSD), γ-hydroxybutyrate, ketamine, heroin, prescription pills (benzodiazepines or antidepressants), cough syrup, Viagra, and nitrous oxide.

U. McCann and G. A. Ricaurte of Johns Hopkins reported in October 1998 that MDMA affected 5-HT neurons. Carbon-11–labeled McN-5652 studies revealed that the availability of 5-HT transporter sites on presynaptic neurons was greater in MDMA users than in controls (McCann and Ricaurte, 1998). It could not be shown whether the effects on serotonergic neurons are permanent or reversible.

This same group of investigators showed that MDMA users who have long abused the drug experience significant memory deficits. Others studies have shown that MDMA users have no changes in pain sensitivity, pain endurance, and pain tolerance compared with control subjects (McCann et al., 1994). Sleep, mood, anxiety, aggression, memory, and appetite are all somewhat affected by MDMA use, possibly related to damage to the 5-HT system. MDMA affects the dopaminergic system, but its primary effect is on serotonergic neurons.

13.3. Lysergic Acid Diethylamine

Among the earliest psychotropic drugs was LSD, synthesized in 1938 by Swiss chemist Dr. Albert Hofmann in Basel, Switzerland, while searching for useful ergot alkaloid derivatives. LSD was widely taken as a hallucinogen by the Mayan-Aztec people of Mexico, and had been used in Greek and Indian Vedic ceremonies. The drug comes from ergot, a fungus growing on rye and other grasses. Hofmann did not recognize its psychedelic properties until April 1943, when he felt the effects on his brain when a tiny amount was accidentally absorbed through his skin. Larger amounts impair perception and motor coordination, causing great anxiety and panic. Despite its harmful side effects, the drug became widely taken by young people in the 1960s.

Hofmann is now 100 years old, and lives in the Alps. At a recent symposium in Basel, Switzerland, he noted that his "problem child," LSD, helps people get closer to nature, even though the drug is widely abused. He admits that the drug can be dangerous and calls its widespread promotion by people, such as Timothy Leary, a crime. "It should be a controlled substance with the same status as morphine."

Timothy Leary, in his *Guide to A Successful Psychedelic Experience*, listed what he believed to be the virtues of LSD: increased personal power, intellectual understanding, sharpened insight into self and culture, improvement of one's life situation, accelerated learning, professional growth, duty, helping others, providing care, rehabilitation, rebirth for one's fellow men, fun, sensuous enjoyment, esthetic pleasure, interpersonal closeness, pure experience, liberation from ego and space-time limits, and attainment of a mystical union. LSD could enable "a person to step out beyond problems of personality, role, and professional status." Unfortunately, many young people believed him, ruining their lives. When he was dying of inoperable prostate cancer, he planned an elaborate death ritual, where his admirers could watch him commit suicide on the Internet. Instead, he died quietly in his sleep in May 1996, surrounded by family and friends. His son Zachary said that his last words were "Why not?" and "Yeah." His mantra had been "turn on, tune in, and drop out."

During the 1950s and early 1960s, before it was illegal in the United States to take LSD, Robert Anton Wilson, who died in 2007, was a prolific American novelist, essayist, philosopher, psychologist, futurologist, and anarchist. He often advocated taking psychedelic drugs (*Sex and Drugs: A Journey Beyond Limits*). It was a "way to break down conditioned associations … to look at the world in a new way … My goal is to try to get people into a state of generalized agnosticism, not agnosticism about God alone, but agnosticism about everything …". A psychedelic dose causes a rush of thoughts, a lot of free association, and some visualization (hallucination) and abreaction (memories so vivid that one seems to relive the experience). A psychedelic dose produces a total but temporary breakdown of how one perceives one's self and the world and produces a "peak experience" or "mystic transcendence of ego." He convinced a lot of young people. The son of a colleague jumped off the roof of his college dormitory, snuffing out what could have been a brilliant and productive life.

In the past, LSD was used as "psychedelic psychotherapy" to treat persons suffering from alcoholism (without success), heroin addicts, and persons with terminal cancer with some success. Unfortunately, LSD causes profound distortions in the perception of reality. People begin to have visions, hear voices, feel bizarre sensations, and manifest rapid, intense emotional swings.

In 1943, it was clear that even small doses of LSD could produce psychotic symptoms by blocking the firing of the neurons that secrete the neurotransmitter 5-HT. These 5-HT neurons are widely distributed throughout the brain and spinal cord, and they are involved in the control of behavioral, perceptual, and regulatory activities of the brain. They affect mood, hunger, body temperature, sexual behavior, muscle control, and sensory perception.

More and more, PET scans of the brains of drug addicts are shown in court. Whenever a new physical or chemical measurement of the human brain is invented, lawyers try to relate the findings to deny free will on the part of their clients. The law assumes that everyone has a free will, and are responsible for their actions. Lawyers try to present evidence that biology drove their clients to criminal actions.

Karl Mann and associates examined opiate receptor availability with C-11 carfentanil PET in patients with a long history of alcoholism. They found an elevation of μ opiate receptors in the striatum and thalamus and certain cortical regions. Blockade of μ opioid receptors by naltrexone reduces the relapse risk among abstinent alcoholics (O'Malley et al., 1992, 1996).

13.4. Nicotine Abuse

Smoking remains a major public health problem. According to the World Health Organization, there are 1.1 billion smokers worldwide. According to the U.S. Center for Disease Control, there are 46.2 million adult cigarette smokers in the United States. Tobacco use in the United States results in >440,000 deaths each year (about one in five deaths.) The economic costs (medical costs and lost productivity) of tobacco use are >$150 billion.

Tobacco contains nicotine, which binds to nicotinic acetylcholine receptors. These receptors bind acetylcholine, but they also are activated by nicotine, an action inhibited by curare. Tobacco is a stimulant that improves alertness, memory, and mood, but it also forms a strong physical and psychological chemical dependence (addiction). Further-more, nicotine can increase anxiety, restlessness, and disturb metabolism. The continual activation of nicotinic acetylcholine receptors can cause a person to become addicted to smoking, requiring progressively increasing doses.

In 2000, public health physicians convinced government officials that guidelines should be issued to get every smoker to quit and to provide medications to help them. Using nicotine patches or gum increases the odds that a person will stop smoking. Behavior can modify brain chemistry.

The "hoodie" skin patch is a cactus-like plant that grows exclusively in South Africa. Several compounds in hoodia suppress appetite. It contains a molecule called P57 that is 10,000 times as active as glucose. It makes one feel full and stunts the appetite. Can-nabinoid CB1 receptor blockade is also effective in reducing nicotine-seeking behavior, supporting the fact that tobacco and marijuana addiction are linked.

On February 29, 1996, retired Army general Barry McCaffrey was sworn in as President Clinton's new director of the Office of National Drug Control Policy, and was given the responsibility for overseeing counter-drug operations in Central and South American countries. Between 80 and 90% of the cocaine consumed in the United States at the time was either produced and/or distributed in Columbia, as well as some 60% of the heroin. His efforts have been only partly successful.

In the *Independent* (May 2007), DeGrandpre tells why he believes the United States is losing its war on cocaine: "America has spent billions battling the drug industry in Bolivia, Colombia and Peru. Production as high as ever, street prices at a low, and the governments of the region are in open revolt." DeGrandpre believes that the distinction between clinical medicines and drugs of addiction is a social construct rather than a physiological fact.

DeGrandpre points out that rural Andean communities have been chewing coca leaves for centuries with no signs of addiction or dependence, no breakdown in traditional social structures, and no apparent ill health. He asks "How did an entire generation of Americans survive the 1920s while consuming prodigious quantities of coca extract in tonics, unguents, pep pills and the most successful product of a world-famous but highly litigious soft drinks manufacturer? How did Ritalin—a drug whose psychoactive effects are generally indistinguishable from those of cocaine in clinical trials, and which is rapidly acquiring high status as a party drug—become one of the most feted and widely prescribed pharmaceuticals of the twenty-first century?"

DeGrandpre's answer is that the pharmacological action of a particular drug cannot be considered in isolation from the context in which it is taken. "Drugs are animated by the ecology of the human settings they enter—psychosocial, cultural and historical—and it is in these powerful and complex settings that drug discourse and so-called drug effects emerge." He cites evidence that neither addiction nor dependence nor the "high" are identifiable as consistent physiological phenomena. All are, he argues, functions of "the ritual of drug use, used in combination with a certain attitude. His central idea is that the widespread use of various drugs has much more to do with culture than chemistry. Today, we are able to study the effects of drugs on the brain and behavior in ways not possible in the past. We can examine molecular processes and the effects of drugs on these processes in different regions of the brain.

The insula, which lies between the temporal and parietal cortex, plays a major role in addiction. Neuronal action potentials from autonomic physiological events play a major role in emotions. Walter Cannon proposed in his book *The Wisdom of the Body* that the autonomic nervous system brings emotions to consciousness. The anterior insula collaborates with posterior insula... and limbic cortex, whereas the posterior insula is involved with auditory, somesthetic, and skeletomotor functions. fMRI indicates that the insula plays a major role in the experience of pain, anger, fear, disgust, happiness, and sadness.

fMRI studies also have shown involvement of the insula in conscious desires, including food and drug craving. The insula integrates all this information about the state of the body with higher order cognitive and emotional processes. The insula governs "homeostatic afferent" sensory pathways via the thalamus and sends output to many other limbic-related structures, such as the amygdala, the ventral striatum, and the orbitofrontal cortex.

Aldous Huxley in his 1932 book *Brave New World* described a fictional mind-altering drug called Soma. "By this time the soma had begun to work. Eyes shone, cheeks were flushed, the inner light of universal benevolence broke out on every face in happy, friendly smiles … Soma may make you lose a few years in time, but think of the enormous, immeasurable durations it can give you out of time. Every soma-holiday is a bit of what our ancestors used to call eternity. People are happy; they get what they want, and they never want what they can't get. They're well off; they're safe; they're never ill; they're not afraid of death; they're blissfully ignorant of passion and old age; they're plagued with no mothers or fathers; they've got no wives, or children, or lovers to feel strongly about; they're so conditioned that they practically can't help behaving as they ought to behave. And if anything should go wrong, there's soma." Is there a soma in our future?

Ephedra was a ritual drink taken by early Indo-Iranians and later adopted by Vedic and Iranian cultures. It is still widely taken in China. In the book *Rigveda*, hymns praise its intoxicating qualities. Aldous Huxley believed that "The purpose of life is not the maintenance of well being, but some intensification and refining of consciousness, some enlargement of knowledge." Soma acted as a tranquillizer, perhaps by increasing the levels of the neurotransmitter, serotonin, in the brain.

Attempts to alter mental activity with alcohol, tobacco, and caffeine go back to ancient times. We have now reached a tipping point when the taking of drugs must become more rational, or civilization will continue along a dangerous path. We are at a tipping point as we continue every day to alter the chemistry of our brains.

Belief in the unending perfectibility of the human species began in the eighteenth century, replacing the pessimistic view of human nature of the sixteenth and seventeenth centuries. Will society increase the use of mind-altering drugs to try to improve human behavior, not just to treat persons with a mental distress?

People believe that the use of performance-enhancing drugs by athletes is a widespread, serious problem. Leon Kass, formerly chairman of the government's Council on Bioethics, agrees, "Let's assume that you found some kind of drug that was safe and had no bad side effects. Let's assume they were legal … I think that baseball players would still be ashamed to be shooting up before they went to bat … it's very much like all kinds of other human activities in which to flourish really means deeds that somehow flow in an uninterrupted way from our souls and bodies."

More and more people in modern society are using behavior-modifying drugs. Ritalin helps correct attention deficit disorder and Prozac relieves depression. Thomas J. Farrell has suggested that there is a vast conspiracy to medicate normal sadness in his review of the book *The Loss of Sadness* by Allan V. Horwitz and Jerome C. Wakefield (August 25, 2007). In our contemporary American culture, people take a pill to make sadness go away, despite the fact that sadness may serve a healthy psychological purpose.

Stimulants and other drugs are often used to improve normal performance, or correct exaggerated aggressiveness or rage. In one of the council's reports, Dr. Kass wrote that "Artificial enhancement can certainly improve a child's abilities and performance, but it does so in a way that separates some element of that achievement from the effort of achieving. It may both rob the child of the edifying features of that effort and teach a child, by parental example, that high performance is to be achieved by artificial means."

An article in *The New York Times* on July 16, 2006, by Jane Gross pointed out that half of all the youngsters in summer camps in the United States take daily prescription medications. Allergy and asthma drugs top the list, but behavior modification and psychiatric medications are "now so common that nurses who dispense them no longer try to avoid stigma by pretending they are vitamins … It's not limited by education level, race, socioeconomics, geography, gender or any of those filters."

About a quarter of the 3 million campers in the 2,600 camps belonging to the American Camp Association are medicated for attention deficit disorder, psychiatric problems, or mood disorders, according to Peg L. Smith, chief executive officer. The proliferation of children on stimulants for attention deficit hyperactivity disorder (ADHD), antidepressants, or antipsychotic drugs is not only in the setting of camps. It is the extension of an increasingly common year-round regimen in schools and at home.

"There is no doubt that kids are more medicated than they used to be," said Dr. Edward A. Walton, a pediatrician at the University of Michigan. A frequent indication is ADHD in children who cannot control their behavior or pay attention in school. They lack self-control, often leading to drug abuse, traffic violations, theft and sexual offenses. They are likely to smoke, abuse alcohol, marijuana, cocaine, and other drugs. Volkow and colleagues at the NIH have found abnormalities in dopaminergic neurotransmission in persons with ADHD. "If you take a drug of abuse, whether it's alcohol, nicotine, cocaine, or amphetamine—it doesn't matter—what you are doing is temporarily increasing the concentration of dopamine in the brain … taking the drug makes them feel better and temporarily perform better."

Questions have been raised about whether all the drugs given by physician's prescription are necessary or even helpful. All of this drug treatment is open to question, but is accepted by large numbers of the public. If we can correlate certain undesirable characteristics, such as unwarranted aggression and violence, with specific neurotransmitter systems, it might be possible to modify undesirable behavior with specific, nonaddicting drugs.

Patients may include psychopaths who express habitual aggression and are responsible for much crime, some of which is violent. Impulsivity, aggression and lack of empathy are often observed in some patients following head trauma or encephalitis, which supports an organic cause. Drugs that block serotonin transporters and increase synaptic serotonin levels help these people. Hormonal effects also are involved, because of gender-related differences in the incidence of violence and aggressiveness. More than

60% of violent crimes committed by women occur during the premenstrual weeks (K. Dalton, *The Premenstrual Syndrome*, Chas. Thomas, 1964).

More and more is being learned about how drugs affect the brain. As we increase our knowledge of the roles of nature and nurture in human behavior, we need to recognize the importance and magnitude of decisions that must be made.

Anthropologist Franz Boas and colleagues Ruth Benedict and Margaret Mead proposed that cultural, rather than biological, factors are the major determinants of personality. They rejected the ideas of those in the 1920s and 1930s who believed in biological determinism. Mead lived with and studied the Manus, the Arapesh, the Mundugumors, and the Balinese people in the South Pacific. She wrote, "We are forced to conclude that human nature is almost unbelievably malleable, responding accurately to contrasting cultural conditions ... Cultures are manmade, they are built of human materials." Biological factors set "limits which must be honestly reckoned with." Howard Gardner quoted Mead as believing that "a single human nature, whatever its limits, could nonetheless spawn a large variety of cultures." Nurture was far more important than nature.

In the search for biochemical correlates in the brain of persons with undesirable human behavior, it is important to distinguish causes from effects. We look for manifestations rather than causes. Abnormalities in brain chemistry are a state of the patient, not necessarily a biochemical trait. We look for antecedent events but do not assume that they are ultimate causes.

Giovanni DeChiro and colleagues at the NIH were the first to use FDG to image brain tumors. Their work derived from that Otto Warburg, born on October 8, 1883, in Freiburg, Germany. He studied chemistry under the great Emil Fischer and then obtained the M.D. degree in Heidelberg in 1911. As a result of the increased anaerobic glucose metabolism in tumors, brain tumors avidly accumulate FDG, increasing with increasing degrees of malignancy. Subsequently, FDG has been used to reflect neuronal activity.

Regional glucose metabolism is decreased in the same areas of the brain where dopaminergic activity is decreased in aging persons, i.e., in those regions of the brain involved in cognition—the frontal regions and the anterior cingulate gyrus. This decline in glucose utilization is associated with deterioration of mental functions, such as problem solving, the ability to think abstractly, and the ability to carry out multiple tasks simultaneously. The anterior cingulate gyrus is related to attention span, impulse control, and mood.

In Parkinson's disease (PD), neurons in the substantia nigra have degenerated. Dopamine transporters located on presynaptic neurons and return dopamine to presynaptic neurons (reuptake). These presynaptic neurons are impaired in PD. The dopamine transporter located on presynaptic dopamine neurons is the mechanism for removing dopamine from synapses, as well as a major site of action for psychostimulants, such as cocaine and amphetamine and for neurotoxins that induce parkinsonism. M. J. Bannon and colleagues showed that the post-mortem content of mRNA encoding the dopamine transporter in the substantia nigra of 18- to 57-year-old subjects was normal, whereas in subjects >57 years of age, there was a precipitous (>95%) decline in substantia nigra dopamine transporter mRNA.

Deprenyl is a selective, irreversible inhibitor of the B-type monoamine oxidase (MAO), a predominantly glial enzyme in the brain. The activity of this enzyme increases with age. Deprenyl inhibits MAOB, which interferes with the reuptake of the monoamines dopamine, 5-HT, and epinephrine. Deprenyl decreases the aging effects on the enzyme

MAOB by blocking entry into neurons of the catecholamines, dopamine, norepinephrine, and epinephrine.

Dopamine levels in the prefrontal cortex decrease together with cognitive ability (Todd S. Braver and Deanna M. Barch of Washington University, *J. Exp. Psychol. Gen.* vol. 130). Dopamine levels in aging adults are associated with difficulties in attention and memory.

To assess the role of serotonin in synaptic neurotransmission, C-11 McN5652 was developed for assessment of serotonin reuptake sites (*Science* 226:1393–1396, 1984). This PET tracer binds to 5-HT transporters. In male subjects, there was a decline of 46% in 5-HT transporter receptors in the caudate nucleus and a 43% decline in the putamen between the ages of 19 and 73. In females, there was a 25% decline between the ages of 19 and 67. Carbon-11 and fluorine-18 altanserin have a high affinity and selectivity for the 5-HT2 receptors. In PET studies with C-11 altanserin, researchers at the University of Pittsburgh found that 5-HT receptors declined by 55% between the ages of 18 and 76. Post-mortem studies also have shown a decrease in serotonin neurons with age.

Serotonin is found not only in the brain but also in storage vesicles within circulating blood platelets. It has both vasodilator and vasoconstrictor effects. If the vascular endothelium is intact, 5-HT released from platelets causes vasodilatation, related to the release of nitric oxide. It can cause vasodilatation by inhibiting the release of norepinephrine. If the vascular endothelium is damaged, 5-HT causes vasoconstriction, and it amplifies the effect of other vasoactive substances, such as histamine, angiotensin II, and norepinephrine. Coronary blood flow is decreased by the release of 5-HT from circulating platelets, especially when the coronary vessels are damaged.

In 1990, Diksic used carbon-11 α methyl-tryptophan to measure the rate of synthesis of serotonin from tryptophan. The rate of accumulation of the carbon-11 tracer can be quantified using a model similar to that used for FDG. With increasing age, glucose metabolism measured with FDG decreases in the frontal regions and the anterior cingulated gyrus. The frontal lobes are involved in cognition and other higher brain functions, whereas the anterior cingulate is involved in attention, impulsiveness, and mood.

13.5. GABA and Glutamate

GABA is the most abundant inhibitory neurotransmitter in the brain. When it binds to postsynaptic receptors, it inhibits the transmission of action potentials from pre- to postsynaptic neurons. A single receiving neuron has thousands of receptor sites and receives many different stimuli at one time. Each neuron either does or does not pass its information along to other neurons. GABA stops the messages from being transmitted. If GABA is insufficient, thousands of neurons send messages rapidly, intensely, and simultaneously, which can result in an epileptic seizure. Enzymes keep GABA concentrations at an optimal level.

Glutamate is the most abundant excitatory neurotransmitter in the brain. By stimulating glutamate receptors, it facilitates synaptic transmission from pre- to postsynaptic neurons. Glutamate is released by the passage of action potentials down a presynaptic neuron. Glutamic acid is involved in learning and memory. Both the pre and postsynaptic neurons have glutamic acid-reuptake systems, which lower synaptic glutamic acid concentration. In excess, glutamic acid triggers excitotoxicity, and it can cause neuronal damage.

In 1992, Dewey and colleagues from BNL studied the effects of a suicide inhibitor of the enzyme that breaks down GABA, an inhibitory neurotransmitter. Inhibition of the enzyme that metabolized GABA resulted in decreased production of GABA. This resulted in decreased inhibition of the dopaminergic neurons, so synaptic dopamine secretion increased.

Dewey and colleagues were able to demonstrate relationships among three neurotransmitters: GABA, dopamine, and acetylcholine. Their studies were another example of measuring endogenous neurotransmitter release by measuring competitive inhibition of radioligand binding.

Recently, the gene GRM3 has been identified that controls secretion of glutamate. Michael Egan, Daniel Weinberger, and colleagues at the NIH believe that mutations of the gene may increase the risk of developing schizophrenia.

Maria A. Oquendo and colleagues from Columbia University (*Neuropsychopharmacology* 30:1163–1172, 2005) used FDG/PET to examine the effect of fenflouramine, which was used to aid in weight reduction. The Food and Drug Administration took it off the market for the treatment of depressed patients. Obesity is one of today's most significant health problems. Fenflouramine (Phen-Fen) was originally hailed as a godsend for those who are grossly overweight.

14

Mentall Ilness

The effect of environmental toxins on mental activity is an important direction of research in molecular imaging. The blood-brain barrier keeps harmful substances from entering the brain from the circulating blood, while transporting nutrients needed for mental functioning. Alcohol readily crosses the blood-brain barrier, affecting the acetylcholine, serotonin, GABA, N-methyl-D-aspartate (NMDA), and glutamate systems. Also affecting mental activity are herbal extracts, such as those from the belladonna plant, *Atrophia belladonna*, which contains atropine and blocks the action of acetylcholine.

Regional cerebral blood flow (rCBF) was measured with positron emission tomography (PET) and oxygen-15 water after administration of 4-methylenedioxymethamphetamine (MDMA) (1.7 mg kg^{-1}) to MDMA-naïve subjects. MDMA is known by the name Ecstasy. The drug produced a bilateral increase in rCBF in the ventromedial frontal-, the inferior temporal-, and the medial occipital-cortex, and the cerebellum. It produced a bilateral decrease in rCBF in the superior temporal cortex, the thalamus, and the preparacentral cortex, in addition to significant decreases in the left amygdala. These regional changes in rCBF paralleled the psychological effects of MDMA, such as mood enhancement and increased sensory perception. (Gamma et al., 2001).

The hallucinogenic drug psilocybin, which comes from mushrooms, causes mind-altering experiences in normal volunteers (Griffiths et al., *Advances in Substance Abuse*, 1980). Psilocybin has hallucinogenic properties closely related to mescaline. Both have been used in tribal rites for centuries by the Aztecs in Mexico. Mescaline-containing plants called magic mushrooms are a schedule 1 substance under the U.S. Controlled Substances Act, putting them in the same category as heroin and lysergic acid diethylamide (LSD). In 2001, the Food and Drug Administration approved a study of the effects of psilocybin in treating patients with obsessive-compulsive disorder, supported by the Multidisciplinary Association for Psychedelic Studies.

Participants reported feelings of intense joy, detachment from reality, and feelings of peace and harmony. In 30% of the volunteers, the drug provoked harrowing experiences of fear and paranoia.

Psilocybin is rapidly dephosphorylated in the body to psilocin, an agonist of serotonin (5-hydroxytryptamine [5-HT]$_{2A}$) receptors. The stimulating of 5-HT receptors by hallucinogenic drugs has given rise to the hypothesis that schizophrenia may be caused by an imbalance in the metabolism of 5-HT. Depressive and catatonic states in schizophrenic patients are thought to be the result of 5-HT deficiency.

In fluorine-18 fluorodeoxyglucose (FDG) studies of the brain of a subject taking psilocybin, there was increased accumulation of the tracer in the frontal cortex and thalamus. Franz Vollenweider at the University of Zurich Psychiatric Hospital reported that psilocybin induces schizophrenia-like psychosis in humans via a 5-HT$_2$ agonist action (*Cogn. Neurosci. Neurorep.* 9(17):3897–3902, December 1, 1998).

Psilocybin and related drugs have possible value in the treatment of "the anguish of impending death," as well as "alcoholism and other forms of drug addiction." (Charles Schuster, Wayne State University, former director of the National Institute on Drug Abuse).

"The misuse of these substances ... cannot be allowed to continue to curtail their use as tools for understanding the neurobiology of human consciousness, self-awareness and their potential as therapeutic agents." Eating psilocybin mushrooms has been popular at raves, clubs and, increasingly, on college campuses, being abused by teenagers and young adults.

Dr. Timothy Leary, Aldous Huxley, Richard Alpert, and others used graduate students as test subjects to study the creativity-enhancing properties of psilocybin. (Later, it was declared a crime for Dr. Leary to carry out such studies.).

Regions of the brain involved in substance abuse are the hippocampus, nucleus accumbens, and amygdala. The reward networks include the dopaminergic neurotransmission system. Addiction to alcohol, heroin, cocaine, and methamphetamine is characterized by a reduction in D$_2$ dopamine receptors in the basal ganglia that are involved in glutamate release. Glutamate is a major excitatory neurotransmitter bound by glutamate receptors. A related molecule, GABA, binds to glutamate receptors, making them less excitable, which brings about the sedative effects of drugs, such as alcohol.

The discovery of a naturally occurring peptide called endorphin in the brain led to a search for pain-killing drugs that would be less addictive than morphine or other opiates used to relieve pain. The new drugs would be as "natural" as endorphin, and they could provide new insight into the molecular processes of addiction.

A most interesting question is, What is the biological role of endorphins? Is their major function to relieve pain or to have a soothing effect? (Thomas, 1983).

W. Schultz and colleagues in Switzerland (*Neuroscientist* 2001 4:293–302) examined the relationship of the dopaminergic system to the rewarding "feel-good" circuits in the brain. They found brief increases of dopaminergic activity after primary food and liquid rewards, as well as following conditioned, reward-promising visual, auditory, and somatosensory stimuli. Neuronal networks in the striatum, orbitofrontal cortex, and amygdala are involved in assessing the quality, quantity, and degree of preference of rewards. The most common drugs producing "feel-good" emotions are alcohol and caffeine.

Thus, the brain can be viewed as involving an unbelievably complex collection of molecules, performing their functions according to the laws of physics and chemistry. The organism functioning as a whole behaves according to the principles of biology. Molecular processes, such as those involving dopamine or 5-HT, must be examined in the context of the biological manifestations of life, such as consciousness, free will, emotions, thoughts and actions. We can now begin to relate emotions and reason to underlying molecular processes.

McHugh and Slavney write, "The concept of disease rests on the proposition that a group of people who are identical in such attributes as symptoms, signs, and laboratory findings may be usefully viewed as sharing some biological abnormality called a disease

... The diagnostic process is an inferential exercise in observation, differentiation and classification ... it allows a more rational basis for prognosis and treatment ... Diseases are the outcomes of biological changes that can be understood as processes of nature ... The disease concept in psychiatry is fundamentally no different from that in general medicine."

Although we strive to define disease on the basis of objective, measurable criteria, we must continually remind ourselves that diseases are not ontologically rigorous; they are not "things" inside the patient that can be identified and, in cases such as cancer, can be removed surgically. In mental illness, making the diagnosis can be harmful by preventing the treatment of the patient as a unique individual.

Molecular medicine provides quantifiable, objective criteria that can be used as the basis for treatment, particularly in the use of drugs specifically designed to correct abnormal biochemical processes. Molecular abnormalities in different parts of the body, including the brain, will become the language of neuropathology. Of course, these therapeutic decisions must involve subjective as well as objective criteria. Rather than search for ultimate causes of the patient's illness, we search for antecedent events or findings. Probability theory, particularly the use of Bayes' theorem, plays an important role in our model of the diagnostic process.

Molecular signatures provide molecular targets for therapy. Increasingly, we focus on patients rather than diseases. Our present is derived from the past. We have found the right road and are moving into the future with a new definition of disease.

"Tipping point," a term introduced by Morton Grodzins, a professor of political science at the University of Chicago, refers to any dramatic moment when something considered of limited applicability becomes commonplace. For example, most white families remain in a neighborhood as long as the number of black families remains small, but, at a certain point, when one too many black families move in, the remaining white families move out.

14.1. Prevention

On September 21, 2005, Elias Zerhouni, Director of the National Institutes of Health (NIH), pointed out that medical science is entering a revolutionary period, shifting its focus from acute to chronic diseases, emphasizing the prevention of disease. Thomas Insell, Director of the National Institute for Mental Health, points out that there are 1,000 different brain diseases, affecting 70 million people in the United States and consuming >6% of the gross national product. More than 50% of incarcerated criminals suffer from mental illnesses, which manifest themselves between the ages of 14 and 21 years. There are 30,000 suicides and 15,000 homicides in the United States every year.

What should be done when a person who has not committed a crime but whose brain scans reveal chemical characteristics that have been correlated with a high risk for antisocial behavior? Should they have counseling at regular intervals? Or should they take medications to modify their brain chemistry? What should be done if they refuse counseling or treatment with drugs?

In the past, proposed antiviolence campaigns have been controversial. Surgeon General Antonio Novello in 1962 called for social and economic interventions: "If we are to succeed in stemming the epidemic of violence, we must first address the social, economic and behavioral causes of violence. We must try to improve living conditions for millions of Americans. We must try to provide the economic, and educational opportunities for our youths that racism and poverty destroy."

Past-Surgeon General C. Everett Koop wrote, "Regarding violence in our society as purely a sociologic matter, or one of law enforcement, has led to unmitigated failure." Rather, he advocated "additional major research on the causes, prevention and cures of violence, which might lead to new medical/public health interventions."

Someday biological factors in mental illness will be treated in a manner analogous to the treatment of infectious diseases. For example, if a young man develops fever and weight loss, lives in slum housing, has a poor diet, and is often being exposed to people with respiratory diseases, he may develop a disease, characterized as pneumonia. He can then be treated immediately if the tubercle bacilli are found in his sputum. These could be considered biomarkers.

Is it possible that someday we will have biomarkers of aggressiveness and violence? To find such biomarkers will require "major research" but the tools—molecular imaging—to carry out this research are now available.

It is helpful to distinguish biomarkers from biomeasurements. The former is used in the context of a specific use of the latter, which can be judged solely on the basis of its sensitivity and specificity in measuring a specific molecular state or process. Quantification is the key for both measurements and markers.

14.2. Informed Consent

Studies of brain chemistry and behavior can be carried out in a collaboration between the criminal justice system and the NIH. No one would become the subject of an experiment without his or her informed consent. Prisoners cannot give truly informed consent while they are in prison, but a prisoner can be well informed, and never be asked to give up his or her rights to participate in experiments. These studies should never be harmful, and might be beneficial. What is learned might be helpful in treatment. Thus, the reward for the volunteer would be the knowledge obtained by the research, and not money being given for participation or by reduction of his sentence.

Violence is a social problem, caused by poverty, racism, and homelessness, but it also may be influenced by biological factors, such as brain chemical changes related to substance abuse. Initial studies might include examining the effect on brain chemistry during the playing of sports, learning to play musical instruments, or studying for college degrees.

"Not long ago, Britain's Drinking Water Inspectorate announced that they'd found Prozac in the country's drinking water. Needless to say, the citizenry took it very calmly. Apparently the drug is ending up in their drinking glasses by way of the water supply. Yum! There's nothing quite as tasty as Prozac—Already Been Calming." Some people may have slowly releasing drugs implanted under their skins to correct specific molecular abnormalities in their brain (*The Mad Dog Weekly*, 2004).

In his novel *1984*, George Orwell described Soma, the ultimate happiness drug. Is marijuana a version of Soma, where sales in the United States every year are worth $32 billion? Many people in Britain take the serotonin-reuptake inhibitor (Prozac). In August

2004, Britain's Environment Agency revealed that Prozac has been found in groundwater and rivers that provide Britain's drinking water.

A "Polypill" was proposed by Nick Wall and Malcolm Law to be taken for the rest of his or her life by everyone over the age of 55, the population accounting for 96% of all deaths from coronary artery disease and strokes. The polypill is a combination of several medications—a statin, an angiotensin-converting enzyme inhibitor, a β-blocker, a thiozide, and aspirin. At the World Congress of Cardiology in Barcelona, the World Heart Foundation president Valentin Fuster announced that a version of the pill would be introduced in Spain by 2009. Investigators in New Zealand and Australia are about to begin trials on alternate versions of the pill.

Is the nicotine patch an effort to modify human mental activity? Will a brain Polypill be developed to treat persons susceptible to substance abuse, such as cocaine? We now know that cocaine and methamphetamine flood the brain with dopamine and other neurotransmitters, producing a "high." Methamphetamine (meth), prescribed for patients with severe attention-deficit hyperactivity disorder or narcolepsy, is widely abused. Crystal meth is the crystalline form of the drug that is smoked. It is difficult to imagine that there will ever be supplementation of a communal water supply, but we should remember that salt is iodized, there is fluoride in water and iron and folic acid are supplements in many foods.

Methamphetamine also acts as a dopaminergic and adrenergic reuptake inhibitor and as a monamine oxidase inhibitor. It results in stimulation of the mesolimbic reward neuronal network, causing euphoria and excitement. The responses of the brain diminish the effects of cocaine or meth, which results in a "craving" to take more and more of the drug to overcome the tolerance and recreate the high. Addiction is the result.

Violence is thought by many to be primarily societal in origin, but we are now beginning to identify biological factors that could eventually be a focus for efforts to prevent or preempt crime. Substance abuse that costs the United States half a trillion dollars a year may result from the stress that society imposes on susceptible individuals. Alcoholism is another cause of violence and preventable deaths. One of four families in the United States has a family member who suffers from alcoholism.

Catholics hold a doctrine of "original sin" that accounts for an inborn tendency toward evil actions. In contrast, Jean-Jacques Rousseau believed in the innate goodness of human beings, who are virtuous and free in their natural state. He believed that political and social forces make people aggressive and violent. "Bohemians" in Paris in the nineteenth century and beatniks and hippies in the twentieth century scorned and resisted societal forces, trying to get young people to return to their virtuous natural states. They took drugs, formed communes and promoted free love to express their "authentic selves." Some parents encouraged their children to follow their "natural" instincts. They believed that emotions should dominate reason; that self-esteem was more important than self-discipline. They viewed history as the corruption of noble savages by their colonial masters, and advocated "breaking the chains of oppression."

Violence is an instinct for survival in all animals, including those who existed before the evolution of humans. As societies evolved, nomads in Asia and Africa terrorized those who chose to live in towns. Hammurabi, who ruled Babylon from 1792 to 1750 BC, believed that vengeance would "bring about the rule of righteousness in the land, destroy the wicked and the evil-doers; so that the strong would not harm the weak." Enemies had to be killed, captured or driven away. Long and bloody religious wars occurred throughout the sixteenth and seventeenth centuries. Karl von Clausewicz (1780–1831) wrote, "War is the extension of politics."

The violence of the Civil War that resulted in the death of 360,000 Northerners and 260,000 Southerners is vividly etched in the minds of all Americans, as are the horrors and suffering in World Wars I and II, the Korean and Vietnam wars. Societal forces and politics have not succeeded in preventing war. Moral forces alone have not been able to prevent violence and war. Is it wild-eyed to think that perhaps one day we may know enough about brain chemistry to make violence and war preventable diseases? Will we be able to do so before violence leads to the disappearance of the human species?

At the end of the 1970s, the belief that human beings were innately good weakened. Widespread aggression, excessive competitiveness, and worldwide conflicts were taken as evidence that many undesirable human characteristics were genetic in origin, evolving during the struggle for survival. Tribal hunter-gatherer societies had to fight to survive. Greek, Roman, and other cultures honored fighters, not pacifists. "Competitiveness is as ancient as the hunt: the same fire that rouses the thrill of pursuit is kept kindled by the joy of victory. There is pleasure in the run, of course, but the high glory is in being the first to cross the line ... the exhilaration of winning is lashed to the rush of discovery" (Kay Redfield Jamison, *Exuberance*, Vintage Books, New York, 2005).

14.3. Rule of Law

In the face of this innate violence, moral and social forces evolved to keep people from killing one another. Humans began to organize tribes. Cultural evolution outweighed biological evolution. Biological characteristics of aggression and violence were no longer the principal factor in the survival of the fittest. Laws were passed to govern tribes, which made it possible for people to live longer and reproduce.

Cognitive scientists describe consciousness as an emergent property of the brain. Neurons, axons, dendrites and molecular processes in different regions of the brain produce a unified perception of the external world that the person senses. The brain stores information in short- and long-term memory, perceives the present, and processes the information in logical ways to yield ideas, thoughts, and emotions. Thus, knowledge and emotions are integral parts of the mind in action.

The concept of original sin has fallen into disrepute. Religion is not needed as the basis of morality. Church pews are empty. Heaven and hell are dead issues, having been banished by the eighteenth century Enlightenment.

Terrorists in the Middle East cried, "God is alive" as they exploded with religious fervor. Even in Europe, the heartland of secularization, people began to ask whether God was really dead. The reason for a resurgence of religion is often fear generated by current events. Abraham taught Jews that they need not be afraid, if they lived and behaved morally. Christianity expanded these ideas, but the Greeks and Romans outlawed religions, and executed or banished Jews and Christians. Unfortunately, when Christians regained power, they had adopted the intolerance of the Romans. This culminated in the crusades from 1095 to 1291. The Crusades tried to recapture Jerusalem and the sacred "Holy Land" from Muslim rule. Heretics were persecuted well into the sixteenth century. The Renaissance and Reformation changed the religious climate of Europe.

In the process of creating nations, human beings brought an end to feudal wars, bring-ing together distant, neighbors, putting up kings over territories, fostering common languages, collecting historical memories of heroes and villains, creating schools, and military service. In his book *From Dawn to Decadence*, Jacques Barzun (Harper Collins Publishers, Inc., 2000) said, "The nineteenth century nation-state became the carrier of civilization." Governments became responsible for national defense, policing the neighborhood, building roads, dispensing justice, delivering mail, and running political and other institutions. Laws prevented violence.

For 2,000 years, the Church has taught that refusal to love one's neighbor is a moral failure. This old-fashioned golden rule is becoming popular again, defining social relations and public obligations. People are having more similar thoughts, feelings, interests and goals. "Love your neighbor as yourself" is being proclaimed in churches, but altruism is not common in many parts of the world. Genocide, slavery, rape, and murder are commonplace.

In the United States in the 6 years after the September 11, 2001, attacks, nearly 100,000 people have been murdered. Bill Bratton, police chief of Los Angeles, has warned of a gathering storm of criminal violence in the United States. The Police Executive Research Forum reported that there are 100,000 instances a year of aggravated assault with a firearm. Over the past 5 years, more than half a million people have been such victims.

In the United States, a black man between the ages of 18 and 24 is more than 8 times as likely as a white man of the same age group to be murdered. Nearly half of the nation's murder victims in 2005 were black, and the number of black men who are slain is on the rise.

More than 2 million Americans are in prison, the highest absolute and per capita rate of incarceration in the world. Black Americans, who account for 12% of the population, constitute half of these prisoners. A tenth of all black men between the ages of 20 and 35 are in jail or prison. Blacks are incarcerated at >8 times the white rate. One in three African Americans in their 1930s have a prison record, as do two thirds of all black male high school dropouts (*The New York Times*, September 30, 2007).

There are several explanations for these statistics: (1) Institutional racism; the power structure is not as helpful to blacks as to whites. Juries are often biased. The justice system focuses on drug offenses more likely to be committed by blacks, who are more exposed to illicit drugs. (2) Health care is less for blacks. (3) Blacks receive excessively long prison sentences. (4) Rehabilitation is not emphasized enough.

One third of the homeless people in the United States are black. Young black men commit a disproportionate number of crimes, especially violent crimes, which are not solely attributable to racism or economic hardships. The rate at which blacks commit homicides is seven times that of whites. Beatings and spouse abuse are more common among blacks, at times leading to spousal homicide. Seventy percent of black babies are born to single mothers, who work outside the home. There is no stigma in having a baby out of wedlock. These lonely young women who cannot find someone they would be willing to marry have babies to have someone love them. *Gangsta rap* degrades women and often, with hateful profanity, proclaim that "rape is a good thing."

The lack of paternal guidance and discipline of young children often results in delinquency. The hunger for a father leads to anger and absent self-esteem. Young males form gangs as parental substitutes, common in segregated, impoverished inner city neighborhoods. Young blacks are stigmatized as "acting white" if they study and behave well in school. Often unable to get jobs because of a poor education, dealing with drugs often becomes the next alternative. These drug dealers then become hardened criminals while in prison, and remain unemployable, unreformable, and unmarriageable. Most return to a life of crime. Their brain chemistry is affected by these harmful social and environmental factors, including drug addiction. Their language and threatening behavior reflect what is happening in their brain. Two parents affect brain development more than one.

Functional brain damage must be treated by greatly increasing efforts at rehabilitation, both physical and medical. They need to feel love and learn how to become confidant. Many have never felt love, but have had many beatings. The system may be a problem, but these men can be helped as individuals. Behavior is affected by social factors, but all problems are not the result of systemic racism. Things do not happen; people make things happen.

Crime today is an epidemic just as infectious diseases were in the past. It was the discovery of infectious organisms that led to the inventions of vaccination and antibiotics that eventually led to the control or disappearance of many infectious diseases in developed countries. Advances in the knowledge of brain chemistry and its effect on behavior may result in analogous benefits in helping eliminate undesirable human behavior.

Specific genes have been correlated with violence, excessive aggressiveness and other undesirable traits, but also with intelligence, creativity, entrepreneurship, competitive drive, and mental stamina. Gender and hormonal factors clearly affect aggressive behavior, which is more of a problem in men than in women. For example, in experiments using PET, aggressive behavior has been correlated with a specific enzyme in the human brain that metabolizes epinephrine, dopamine, and 5-HT.

The human brain is no longer a black box; functional magnetic resonance imaging (fMRI) and computed tomography (CT) can define anatomical structures; PET and single photon emission computed tomography (SPECT) can examine molecular processes within the brain and relate them to behavior. fMRI can identify regions of the brain and then PET can examine neuronal activity. For example, fMRI, together with PET, can identify specific regions of the brain that are correlated with vision and speech. Such studies can identify those parts of the brain that are diseased or damaged by trauma.

One can try to relate brain chemical factors associated with destructive behavior, with altruism as well as violence, with cooperability as well as with harmful competitiveness. We have some knowledge of the changes in brain chemistry associated with feelings of well being, happiness, depression, rage, and aggression as well as the molecular processes in the brain related to suicide, homicide, and wars. We can show changes in chemical processes in the brain when a person runs 100 yards or remembers a series of numbers. We can characterize the chemistry in the brain of a person unable to control his or her impulses. We can define the molecular phenotypes of creativity, as well as violence, and examine the relative importance of nature vs. nurture.

Richard Leakey argued that human beings are not innately aggressive, but cooperative and that human behavior is determined chiefly by learning (Darwin and Norton, 1979).

Leakey claimed that Konrad Lorenz and Desmond Morris promoted "the myth that the human species carries with it an inescapable legacy of territoriality, and aggression, instincts which must be ventilated lest they spill over in ugly fashion."

"Among the many phylogenetically adapted norms of human social behavior, there is hardly one that does not need to be controlled and kept on a leash by responsible morality ... the fate of humanity hangs on the question whether or not responsible morality will be able to cope with its rapidly growing burden." (Konrad Lorenz).

"If only there were evil people insidiously committing evil deeds, and it were only necessary to separate them from the rest of us and destroy them ... The line dividing good and evil cuts through the heart of every human being." (Aleksandra Solzhenitsyn, *The Gulag Archipelago*).

According to Christian theology, humans are born in a condition of sinfulness, called "original sin, which was first committed when Adam and Eve succumbed to the serpent's temptation, known as the "Fall." The Jewish religion teaches that humans are born pure. We all make mistakes as we go through life, but there is innate holiness in our lives. Orthodox Judaism teaches that rules of behavior are taught in the laws of the Torah, whereas more liberal Jews are willing to depart from the Torah, but believe that the spirit of the text survives. Today's ethical principles can be traced to their sources in these ancient texts. In the first century, the Jewish sage, Hillel, said, "Do not do to others what you would not want them to do to you. All the rest is commentary, Go and learn it. (B.Shah. 31a). Christians take the Ten Commandments as the cornerstone of an ethical life. Jesus said, "You shall love your neighbor as yourself (Mk,12:28-33). The command to love one's enemy defuses evil and breaks the cycle of violence and war.

In the book *Irreconcilable Differences*, David Sandmel, Rosanne Catalano, and Christopher Leighton (Westview Press, 2001) write that Christians believe that loving one's enemies is justified by God's love for use, a love that knows no limit, and embraces the just and unjust alike. By reading the Bible, Jews and Christians make sense of the world around them. Scriptures bring Jews and Christians together. They discern their "lifestyle" in the Bible, which provides a moral and spiritual guidebook.

Protestants and Roman Catholics wrote each others epitaphs in blood, until the Enlightenment recognized the evils resulting from religious passions. Spinoza wrote that theologians "wring their inventions and sayings out of the sacred text and fortify them with Divine authority ... compelling others to think as they do." (Tractatus Theologico-Politicus, ch. 7, 1). He anticipated the separation of religion from political life, as well as the tolerance of religious differences. The Enlightenment provided important new insights into the words of religious texts to restore valuable principles without the harmful interpretations.

14.4. Free Will

In his classic novel *Crime and Punishment*, Rodion Raskolnikov wrote, "The first manifestations of newly-freed man are exaggerated individualism, self-isolation, and rebellion against the exterior harmony of the world; he develops an unhealthy self-love which moves him to explore the lower reaches of his being" (Foedor Dostoevsky). He believed that freedom opens the path of evil. In all of Dostoevsky's novels, man progresses from freedom to evil, and then on to redemption (Nochlas Berdyaev, *The Nature of Man and Evil*, 1964).

The existence of free will has been debated for >2 millenia. Religions teach that human beings have free will and choose their actions from among various alternatives. They connect free will to moral responsibility. Free will necessitates that a person is responsible for his or her actions, but most people agree that biological processes beyond our control determine our actions. Molecular imaging can help correlate these biological factors with behavior.

14.5. Religious Intolerance

Extremists have carried out terrorist attacks over centuries, frequently the result of religious differences. For example, citizens of Saudi Arabia, Egypt, and Jordan are mostly Sunni, whereas those of Iraq, Lebanon, and Pakistan are mostly Shiites. The schism between Sunnis and Shiite factions dates back to the seventh century B.C. when Abu Bakr, the first Caliph, was chosen as the successor to Mohammed, Islam's founder, rather than his son-in-law Ali. Supporters of Ali sided with Mohammed's son Hussein. Shiites make up only 10% of Islam's 1.5 billion persons throughout the world, and constitute 60% of the population of Iraq. Islamic radicalism in the Arab world is related to a failure of several nations to advance economic growth, or political and cultural freedom. Disaffected youth justify their violent behavior, including murder and suicide of civilians, by quoting the Koran:

"Fight in the way of Allah with those who fight against you... God does not love the transgressors. Kill them wherever you find them and drive them out [of the place] from which they drove you out and [remember] persecution is worse than carnage. But do not initiate war with them near the Holy Kabah unless they attack you there. But if they attack you, put them to the sword [without any hesitation]. Thus shall such disbelievers be rewarded. Keep fighting against them, until persecution does not remain and Allah's religion reigns supreme. But if they mend their ways, then [you should know that] an offensive is only allowed against the evil-doers."

Islamic Sunnis and Shiites in the Middle East have experienced violence since the death of the Prophet Muhammad in 622 A.D. The Sunni–Shia division has been a raw wound on the body of Islam since the eighth century when Moslem armies defeated those of the Persian Empire, an empire that stretched from Iberia to the Indus Valley. They manifest a fundamentalist hostility to Christianity, the other sect of the Moslem religion and the values of freedom and democracy.

In the seventh century, Mohammed ordered the massacre of Jews whom he accused of breaking an agreement. Hundreds were beheaded, and their bodies thrown into open trenches. According to Tariq Ramadon, Mohammed was a lovable person, transcending his ego. His love of God freed him from depending on other human beings (*Islam, The West and the Challenge of Modernity*, Leicester Press).

During the 1960s, 15 million Muslims immigrated to Western Europe from North Africa to France, and Spain; from South Asia to Britain; from East Asia to Holland; and from Turkey to Germany. They faced permissive, consumer-oriented cultures, which drove them to re-emphasize the rules of the Koran. The young in particular asserted their Muslim identities. Separatist groups advocated leaving mainstream society. Second and third generations felt marginalized by racial and ethnic discrimination, and religious hostility.

On September 11, 2001, 19 young suicidal terrorists crashed two commercial jets into the World Trade Center in New York in the name of Allah. They were all from an educated upper class, and they viewed themselves as revolutionaries searching for justice.

On July 7, 2005, Muslims born and educated in Britain carried out the attacks on the London transport system that killed 56 people and injured >70 persons.

Most Muslims living in Western Europe denounce violence and condemn the attacks that killed women and children. The terrorists were young men lacking jobs, poorly educated, and facing discrimination.

Osama bin Laden has characterized suicidal terrorists as helpless and hopeless persons, who murder those who humiliate them. Their salvation lies in hatred and violence. They say: "I am not afraid to kill and die ... I will not be humiliated." They do not believe they are committing sins. They believe that murder is an understandable moral reaction to societal ills. They are similar in many respects to the religious zealots in the fourteenth century Caliphate whose goal was a global theocracy.

In May 2007, President Mahmoud Ahmadinejad of Iran wrote, "Liberalism and Western-style democracy have not been able to realize the ideals of humanity. Today, these concepts have failed. Those with insight can already hear the sounds of the shattering and fall of the ideology and thoughts of liberal democratic systems ... the world is gravitating towards faith in the Almighty and justice and the will of God will prevail over all things" (Letter to President George Bush in May, 2007).

The Moslem Council of Britain said that it is the "Islamic duty" for Moslems to support the efforts of the Council to fight terrorism. Moslems can modernize their societies without succumbing to the dehumanizing forces of secularism. The Sunni–Shia division remains a raw, untreated wound on the body of Islam. Moslem leaders have been criticized for not being more vociferous in denouncing violence and sectarian strife. Killings are contrary to the Prophet's teaching.

To escape a natural tendency to ever-present violence, warfare, reprisals, and vengeance, people must give up some freedoms. Laws are needed to protect human life, freedom, and dignity. Persons who choose to stay outside the law have their own definition of freedom, not believing that law-abiding citizens can retain most of their freedoms to live in peace in the face of violence, war, fear, poverty, loneliness, barbarity, ignorance, and savagery. They want to live in a world of reason, peace, security, and goodwill toward their fellow humans. Sadly, we see in the news every day what humans can do to each other when no societal forces restrain them.

"Social conflict is an inevitability in human history, and will be to its very end" (Reinhold Niebuhr). Sigmund Freud was more optimistic, and hoped that someday it might be possible to understand the biology of the brain. War could be prevented, people would be smarter, able to control their instincts and be less aggressive. Their actions would be determined by reason and the rule of law, rather than by the innate instincts of fear, terror, and war.

Breakdown of government results in assaults and violence in the home, the office, and on city streets. Love of one's children is no longer universal. Killers, drug addicts, and hardened criminals are overwhelming our prisons, which themselves have become scenes of perpetual violence. Gangs of inmates terrify and sexually abuse other prison-ers, overawe guards and often escape. Riots are frequent.

More than 2 million Americans are in prison. Black men in their early 1930s are impris-oned at 7 times the rate of whites in the same age group (Jason De Parte, *The American Prison Nightmare. The New York Review*, April 12, 2004). Whites with only a high school education get imprisoned 20 times as often as those with college degrees. The prison and jail population increased from 380,000 in 1975 to 2.2 million in 2004. The mentally ill account for 16% of the prison population, i.e., about 350,000 on a given day.

More than half of state and federal inmates have been convicted of nonviolent crimes, such as selling drugs. By the 1990s, 60% of federal inmates were in for drug offenses. Spending time in prison can turn minor offenders into hardened criminals. Seven Americans are behind bars for every 1,000 citizens. By the time they reach their mid-1930s, 60% of black high school dropouts are prisoners or ex-convicts. Blacks are twice as likely as whites to be unemployed and go to prison 8 times as often. Prison also is related to poverty. From 1980 to 2000, the number of children with fathers behind bars rose sixfold to 2.1 million children. Among white children, just >1% have incarcerated fathers, whereas among black children the figure is 10%.

Today, armed guards patrol the corridors of many public schools. Danger of assault has become a risk of teaching. Government has not been able to solve these problems. Human beings have an innate sense of moral responsibility, whereas others believe that morality is the result of religious and other cultural factors. Jesus said, "Blessed are the peacemakers," but religions, including Christianity, are thought to have played a major role in human violence and wars.

A succession of recent books have attacked religion—Sam Harris' *The End of Faith*; Richard Dawkins' *The God Delusion*; and Christopher Hitchens' *God Is Not Great–How Religion Poisons Everything*. They deplore the fact that, in their view, science and democracy have failed to kill religion. For much of the twentieth century, religion was banished from politics, but today the drift to secular humanism has been slowed markedly and religions are on the increase, despite that only 20% of Europeans say that God plays an important role in their lives, compared with 60% of Americans. Forty percent of Americans go to church every week. Half of the world's 13 million Mormons live in the United States, and a Mormon, Mitt Romney, is a candidate for the presidency.

In Arabia, fundamentalism is growing as resistance develops to the spread of Western culture and power. Religion is no longer based primarily on inheritance, but is based on adult's making the choice to go to a mosque, synagogue, temple, or church.

Religion has long been blamed for promoting violence and war. For example, we are reminded that in 1095, Pope Urban II sent Crusaders to reclaim the Holy Land from Muslims and "destroy that vile race from the lands of our friends … Christ commands it." Yet, George Weigel, a leading American conservative, has called the twentieth century the most secular and the bloodiest in human history. The Israel–Palestine conflict has been a territorial, rather than a religious, dispute for >40 years. Timothy Shah of the Council on Foreign Relations argues that >30 of 80 countries have become freer in the years between 1972 and 2000 because of religious influences. Established religions pushed for democracy, e.g., the Catholic Church in Poland. Religious people sought freedom, often to be able to pursue their faith.

Only a tiny number of angry Muslims with AK 47s, rocket-propelled grenades, and exploding vests, out of millions of others in the Middle East, Algeria, Somalia, and Thailand, fanatical Muslims believe that they are fighting decadent, evil unbelievers, and enemies of Islam. In Iraq, suicide bombers from radical Sunni and Shiite Muslim groups are considered martyrs and they are not thought to violate the Koran ban on killing oneself. They believe that death in battle against infidels brings great honor. The killing of Shiites by Sunni militants is justified by the belief that they are not true Muslims.

"Religion impels murderous fanatics to acts of terrorism; extreme religious sects act as breeding grounds for young terrorists and murderers" (*Breaking the Spell: Religion as a Natural Phenomenon*, Daniel C. Dennett, Viking Penguin). The wars between India and Pakistan in 1947, 1965, and 1971 were not explicitly about religion but about the disputed territory of Kashmir.

The jihadists' war is not against the West but against apostate Muslim regimes, or occupying powers, such as the Americans in Iraq. Michael Moss and Squad Mekhennet have cited the jihadi tenets of militant Islam (*The New York Times*, June 10, 2007). All relate to human emotions, rather than fanatical religious beliefs:

"You can kill bystanders without feeling guilt ..."Nor take life, which Allah has made sacred, except for just cause ... Fight in the cause of Allah with those who fight with you, but do not transgress limits; for Allah loveth not transgressors ... and turn them out from where they have turned you out, for tumult and oppression are worse than slaughter ..." (Verses 33, 190 and 191, Ch. 17 of the Holy Koran).

"God will identify those who deserve to die and send them to hell. The innocent won't suffer."

"You can kill children without needing to feel distress...if they are killed in a jihad, they go straight to heaven."

"You can single out civilians for killing ... Some government officials are fair game ... It would be legitimate to attack banks because they charge interest, a violation of Islamic law."

"You cannot kill in the country where you reside unless you were born there." (In the September 11, 2001, attacks in New York, the hijackers came from outside the United States.

"You can lie or hide your religion if you do this for jihad."

"You may need to ask your parents for their consent. It is every Moslem's duty to fight the Americans, so it is not difficult to get permission."

Reflecting the views of the majority of Muslims, in July 2007, an impressive re-election victory was won by Turkey's conservative Muslim ruling party. This is an inspiration for those who believe in the cause of democracy in the Muslim world.

The success of the ruling party in Turkey in government shows that political parties grounded in Islam can thrive within a democratic political system. The president is a devout Muslim, but he has made no move during 5 years in office to islamicize the Turkish government or curb the rights of secular Turks. On the contrary, he has pushed through liberalizing reforms, including greater rights for women; presided over an economic boom driven by foreign trade and investment; and pressed for Turkish entry into the European Union.

Voters in Turkey recently rejected the claim of the traditional military–secular establishment that there is any fundamental incompatibility between democracy and Islam.

Instead, they rewarded a party that has given the country its most competent and successful government in recent decades.

In contrast, in Iran in 1979, a Shi'ite Muslim, Ayatollah Khomeini, actively supported international terrorism. His government instituted a strict regime of Islamic law, including the stoning of adulterers and amputating the hands of thieves. In November 1979, student radicals overran the U.S. embassy, took hostages, who were not released until January 20, 1981, the day of President Reagan's inauguration.

The terrorist group Al Qaeda was formed in the mid-1990s, under the leadership of Osama bin Laden. Radical Muslims became terrorists and murderers, carrying out suicidal bombings with a sense of righteousness. They abandoned any feelings they had of moral responsibility inhibiting innate aggressiveness. Their fanaticism also inhibited any culturally evolved morality that normally inhibits aggressiveness. Forces in their brains countermanded their instinctive and cultural inhibitions against killing fellow humans.

In Saudi Arabia, a sect of Islam called Wahhabism, started by Mohammed ibn Abd al-Wahhab, has inspired Islamic extremism since the eighteenth century. Some believe that there is today a clash between Christianity, Judaism, and Islam. Others such as Resa Aslan challenge that idea (*No God But God*, Random House, 2007). Islam is a religion firmly rooted in the traditions of the Jewish and Christian scriptures. The legal code developed by Mohammed to regulate every aspect of Islamic life was in harmony with ideals of democracy and respected human rights. Aslan rejects the idea that the principal cause of terrorism is the Islamic religion. Between 2004 and 2006, the number of Iraqis who supported the idea of an Islamic state fell from 30 to 22% (University of Michigan's Institute for Social Research). Most Iraqis put their national over their Islamic identity. Muslims are the principal targets of terrorism. On January 11, 2007, National Intelligence Director John Negroponte testified before the U.S. Senate Intelligence Committee that al Queda leaders hide out in Pakistan, where they are rebuilding their terrorist network and developing weapons of mass destruction.

The greatest fear is that rogue nations, such as Iran, will develop nuclear weapons. The Iranian government has refused to halt its uranium-enrichment program. The world is on the precipice of a new and dangerous nuclear era. Terrorists not controlled by any nation cannot be deterred from the mass destruction and deaths of hundreds of millions of people. The principle of "mutual destruction" is no longer a force promoting peace. A *Moslem bomb* already exists in Pakistan, a country which is moving toward becoming an Islamic Republic. If Iran develops nuclear weapons, the whole Middle East may explode as Israel and other countries become involved.

Chemical attacks are also a threat, with the threat of attacks on some of the 700 chemical plants in the United States. Such attacks could result in the death of hundred of thousands of people, because heavier-than-air chemical toxins can travel 10–20 miles. Chemical processes are being redesigned to reduce hazards. For example, hydrofluoric acid is being replaced with sulfuric acid, not nearly as dangerous. Every year 100,000 tank cars containing toxic gases such as chlorine or anhydrous ammonia are shipped around the United States. These tank cars are easily identified as they move through densely populated areas. There is also the risk from biological agents, such as smallpox or botulism in milk and other foods.

When microorganisms were identified as causes of infectious diseases, scientists directed their efforts to invent vaccinations and antibiotics Someday, we may be able to diagnose abnormal brain chemistry before negligent, hostile, or criminal behavior follows, before aggressiveness leads to violence, physical injury, or crime.

Men are more aggressive and more likely to be violent than women, related to both biological and cultural factors. In the United States, men are 4 times more likely to commit suicide than women. Women make more attempts, but most are gestures. Violence against women by their spouses or live-in partners is widespread in both rural and urban areas. Nearly 25,000 women at 15 sites in 10 countries are victims of partner violence ranging from a low of 15% in Yokahama, Japan, to a high of 71% in Ethiopia. Only a tiny fraction of spouse abuse is reported to authorities (Rosenthal, 2006).

The differences between men and women are genetic, hormonal, and cultural. Beginning in utero, the male brain needs testosterone to organize it into specific male networks to produce male behavior. It stays a female brain if testosterone is lacking at the proper time of brain development. Once the brain establishes a male or female pattern, the intervention of opposite-sex hormones does not affect it.

Anne Moir and David Jessel, in their book *Brain Sex* (Carol Publishing Group, 1991) cite questionnaires that show that men characterize themselves as being practical, shrewd, assertive, dominating, competitive, critical, and self-controlled, whereas women describe their ideal selves as loving, affectionate, impulsive, sympathetic, and generous. Men value competition, science, power, prestige, dominance, and freedom, whereas women value personal relationships and security. Anger is associated with high blood testosterone levels. In studies of Vietnam veterans, James Dabbs found that men with very high blood levels of testosterone were twice as likely to engage in antisocial behavior such as drug abuse, trouble with the law, and fighting. High testosterone increases the efficiency with which particularly male attributes are performed, and the amount of estrogen influences the success of usual female aptitudes.

Many who commit suicide have a past history of reacting to stress with excessive anger or aggression. The young men who died in suicidal missions or as kamikaze pilots in the closing months of World War II had a profoundly tragic view of life. They were far from the brainwashed zombies that most Americans imagined at the time. The pilots were thoughtful, sensitive men, neither religious nor nationalistic fanatics. They enlisted in a secret brotherhood that gave meaning and purpose to their lives, and were motivated more by loyalty to their comrades than by hatred of the enemy. "Once the operation had been conceived and ordered, it would have been unthinkable and shameful not to carry it out" (*Kamikaze Diaries: Reflections of Japanese Student Soldiers*, Emiko Ohnuki-Tierney, University of Chicago Press, 2006). Like the ordinary Japanese soldier, they were trained to jump up at the mere mention of the emperor, to think of foreigners as devils, and to exalt death as the highest honor.

People rarely ask, "Why do I think this way? Why do I act this way? Do I control my destiny: Or do genes, hormones, psychology, societal, and environmental forces program me? Frustration can lead to depression that can then progress to aggression. They know that pneumonia affects the lungs, but they do not know that violence affects their frontal lobes. They believe that their problem is that bureaucrats organize their lives and limit what they can do. Government, private institutions, universities, corporations, and industries overwhelm them with rules, at times making them depressed or even violent.

In 1848, an iron bar accidentally passed through the skull of a man named Phineas Gage, damaging his frontal lobes. He suddenly developed impulsive and aggressive behavior. His friends said he was "no longer Gage." This provided the first evidence that damage to the frontal lobes could alter personality and affect socially appropriate behavior. It led to the idea that there was localization of emotion and personality in particular parts of the brain. Children who injure their prefrontal cortex have been found to be unable to control their frustration, anger, and aggression.

Of the >3,300 people in prison awaiting execution in the United States, 10% suffer from serious mental illness. In the overall prison population, 17% have serious mental illness. Many become normal when treated with psychoactive drugs. Since the closing of state-run mental health hospitals in the United States a generation ago, 330,000 of the 2.2 million persons in the nation's prisons are mentally ill. Some newly released people become violent when they receive little help finding jobs, housing, and treatment for their mental illness. For example, in Kansas 65% of the admissions to state prisons were due to violations of parole, usually by people with drug addictions or mental illness.

We need to increase our efforts to keep people released from prison to be able to avoid further criminal activity. Well-designed scientific studies based on molecular imaging with fMRI, PET/CT, and SPECT/CT can be designed to try to understand what happens in the brain during specific desirable and undesirable activities.

An increasing cause of mental dysfunction among soldiers is posttraumatic stress disorder (PTSD), which is becoming more common as a result of the war in Iraq. Of 1.4 million soldiers, 13% have developed PTSD, compared with 20–30% who developed PTSD after the Vietnam War. It is clear that the brain retains images of violent events that can emerge as flashbacks years after the traumatic experiences.

Twenty percent of U.S. combat troops in Iraq have PTSD, anxiety, or depression. With PTSD, small stimuli can unleash a flood of painful memories, which, with nightmares, can lead to jumpiness, anger, depression, and intense stress.

"In combat situations, it's appropriate to have an increased fear response," said Emory University professor J. Douglas Bremner. "PTSD is a failure to learn how to turn that off." Bremner and his group were the first to apply brain imaging to the study of PTSD. They used MRI and found that the hippocampus was smaller in patients with PTSD related to combat. Neuropsychological testing revealed deficits in verbal declarative memory function in PTSD. PET showed a decrease in medial prefrontal and anterior cingulate function when symptoms of PTSD were induced.

Persons suffering severe trauma can be helped by the administration of the β-blocker drug propranolol. Patients with PTSD recovered faster than did a control group who received only psychological counseling. None of the persons treated with β receptor-blocking drugs manifest elevated physiological responses when asked to recall their traumatic experiences, whereas 40% of the control group members did (Flemine, October 1, 2006).

14.6. Ethics of Human Brain Research

Well-designed and well-controlled research studies with molecular imaging must be carried out under strict ethical regulations and supervision. Never again will studies such as the Public Health Service Syphilis Study from 1932 and 1972 be performed. In this study, half of 399 black men were given antibiotic treatment for syphilis, whereas others served as controls. When the press exposed this study, it was stopped immediately.

This unfortunate experience led to a National Commission being established to develop the basic ethical principles governing research involving human subjects. Guidelines were developed to assure that human research is conducted in accordance with those principles. Federal regulations protect human research subjects under the National Research Act, 45 Code of Federal Regulations 46, and Title of Code of Federal Regulations 50.

The use of mind-altering drugs has long been and remains controversial. A striking example was the public's response in August 2004 when trace amounts of fluoxetine hydrochloride (Prozac), a selective serotonin reuptake inhibitor were found in drinking water in the United Kingdom. It is not known how it got there. The emotional reaction was even greater than the response of many persons to the first use of widespread fluoridation of drinking water to prevent caries.

For millennia, humans have been attracted to the idea of altering their mental state. Most abused drugs decrease overall neuronal activity in the brain as indicated by a decrease in the overall accumulation of F-18 deoxyglucose. Perhaps, when the brain is at rest, background neuronal activity tends to be unpleasant.

Many American children are being prescribed antipsychotic drugs for attention deficit disorder or other behavioral problems. Drugs such as Zyprexa and Risperdal are often sold for off-label use. What is needed is scientific documentation of the value of these uses at the molecular as well as psychological levels.

15

The Hormone of Love

In William Shakespeare's day (1564–1616), there was wrangling about the relative importance of the heart or the brain as the seat of intelligence and emotions. He wrote, "Tell me where is fancy bred; in the heart or in the head."

In 2007, in her book *The Chemistry of Love*, Teresa Pitman wrote, "When my first baby was born, the doctor laid his wet, pink body on my stomach. As I helped him towards my breast, the room seemed to fill up with light and I felt a new burst of energy. I was overwhelmed by the intense love I felt for this nine-pound boy with his squished-up face and bald head. It was the most powerful emotion I'd ever felt ... your body has been preparing you to fall in love with your baby for a long time–starting long before you became pregnant ..." By age 12, or around puberty, girls begin to prefer to look at baby faces rather than at the faces of adults or people their own age.

Oxytocin and vasopressin were discovered, isolated, and synthesized by Vincent du Vigneaud in 1953, for which he received the Nobel Prize in Chemistry in 1955. Oxytocin has been called the *hormone of love*. It is synthesized in the paraventricular and supraoptic nuclei of the hypothalamus and stored in the posterior pituitary, until released into the blood. Oxytocinergic neurons have widespread projections throughout the central nervous system. Mothers and babies are affected greatly by oxytocin during pregnancy, during the birth process, and during suckling (Odent 2001).

Oxytocin, as well as dopamine, norepinephrine, androgens, estrogens, and vasopressin, is related to the sex drive and sexual arousal, is secreted during sex, and is associated with feelings of love, altruism, warmth, calmness, bonding, tenderness, togetherness, and bodily contact.

Secretion of oxytocin is regulated by the electrical activity of the neurons in the hypothalamus, which generate action potentials that propagate down axons to the nerve endings in the pituitary. These axons contain oxytocin-containing vesicles, which are released when the nerve terminals are depolarized.

Serotonin also affects the emotions of love and romance. Androgens, estrogens, oxytocin, and vasopressin increase the sex drive, whereas increased levels of synaptic serotonin, such as result from the taking of reuptake transporter inhibitors, suppress the sex drive and inhibit feelings of attraction by the opposite sex. Dopamine and norepinephrine increase the sex drive.

Helen Fisher of Rutgers University wrote, "Everything that happens with romantic love has a chemical basis." She reports studies using functional magnetic resonance imaging (fMRI) that show that looking at a picture of one's lover activates the basal ganglia, which she interprets as indicating involvement of the dopamine system.

"Understanding the brain patterns of the newly in love can teach us how to rekindle romance and boost the health of long-term relationships" (*Wall St Journal*, Feb. 13, 2007). "We know that novelty and new experiences engage the dopamine system, and when it is associated with your partner, it creates a link with your partner" (Arthur Aron, Stony Brook University).

In fMRI studies, Helen Fisher found that lovemaking activates areas in the brain involved in dopamine production. Dietary factors also matter. A tryptophan-free diet results in a fall in brain serotonin levels within about 4 h.

Oxytocin is synthesized in peripheral tissues, including the uterus, placenta, amnion, corpus luteum, testes, and heart as well as the brain. Oxytocin receptors are on neurons in many parts of the brain and spinal cord, including the amygdala, ventromedial hypothalamus, septum, and brain stem.

Oxytocin receptors in a high-affinity state require both magnesium and cholesterol as modulators. The function and regulation of the oxytocin system is also steroid dependent. Oxytocin stimulates contraction of uterine smooth muscles during labor and milk ejection during lactation. At the onset of labor, uterine sensitivity to oxytocin increases together with up-regulation of oxytocin receptors in the myometrium. The hormone also plays an important role in other reproduction-related functions, such as control of the estrous cycle length, follicle luteinization in the ovary, and ovarian steroidogenesis. Oxytocin receptors are found in other tissues, including the kidney, heart, thymus, pancreas, and adipocytes.

High oxytocin levels and prolactin stimulate the breasts to make milk. Oxytocin also produces feelings of relaxation and nurturing behavior toward the baby.

In response to the pain of contractions, women's bodies produce endorphins, which reduce pain and promote a feeling of well being. As the baby is born, the contractions stop, but endorphins still flow through the mother's body (and the baby's). The mother experiences a wonderful, euphoric feeling as her baby is placed in her arms.

Stress inhibits oxytocin release, whereas inhaled oxytocin has been reported to reduce fear. The drug MDMA (ecstasy) may increase feelings of love, empathy, and affection for others by stimulating oxytocin activity. Oxytocin secretion is reduced by catecholamines released from the adrenal gland in response to stress. Circulating levels of sex steroids modulate the production of and response to oxytocin. Oxytocin receptors in the uterus are increased late in gestation, as a result of increasing concentrations of circulating estrogen.

Vasopressin may have an effect on the brain, as well as its peripheral actions in controlling urine, blood pressure, and stimulating adrenocorticotropin secretion. Vasopressin seems to have an opposite action to oxytocin, contributing to aggressive behavior.

Serotonin also seems to have an opposite effect from oxytocin. Increasing the levels of serotonin by the administration of antidepressants diminish the sex drive and feelings of attraction to the opposite sex. Patients taking these drugs that increase serotonin levels have an inability to cry, worry, become angry, or be concerned about the feelings of others.

The enzyme monoamine oxidase (MAO) A is an inhibitor of the MAO, that breaks down dopamine, serotonin and epinephrine. MAO levels rise with age, which may account for the fall off in monoamine neurotransmitter activity with age. Inhibition of MAO activity by drugs can prevent this decrease.

16

The Golden Rule

In 1987, Tereo Hiruma, President of Hamamatsu Photonics, visited Johns Hopkins. We learned that we were both impressed by the new ability to measure regional brain chemistry in the living human brain by positron emission tomography (PET). We agreed to start a Japanese and American Program, called *Peace Through Mind/Brain Science*. We agreed to begin a series of annual seminars in Japan in which physicians, scientists, and engineers from all over the world who were interested in trying to better understand brain chemistry, and related pleasure, pain, aggression, violence, rewards, and other emotions that might help advance world peace. The Cold War was still going on at that time.

Among the major accomplishments of the program was cooperating in the creation of the first PET Center in the Soviet Union at Leningrad, as well as beginning the development of a whole series of new and improved animal and human PET and fluorescent (optical) scanners. The meetings emphasized molecular imaging and genetics.

The genetics part the program was guided by Sydney Brenner, who subsequently won the 2002 Nobel Prize in physiology and medicine. Brenner showed that each three-nucleotide sequence of nucleotides in DNA encoded an amino acid. The nucleotide bases in DNA program the sequence of amino acids in peptides and proteins.

At the same time, the Nobel Prize in physics was awarded to Professor M. Koshiba. He had worked with Hamamatsu Photonics in developing photomultiplier tubes that were 25 inches in diameter and then used them in 1979 to measure neutrinos emitted during the explosion of stars. His work led to the creation of neutrino astrophysics.

16.1. Good vs Evil

G. K. Chesterton described his sitting by a pond on a lovely summer afternoon watching children torturing a cat. This convinced him of the reality of original sin (Wills, 1999). Judaism, Christianity, and Islam trace their origins back to the creation of evil by Adam and Eve in the Garden of Eden. St. Augustine taught that God created nature. Adam caused things to go wrong. His fall gave birth to seven deadly sins: lust, gluttony, greed, sloth, wrath, envy and pride. Human virtues persisted: chastity, abstinence, liberality, diligence, patience, kindness, and humility. These virtues provided the foundation of morality and ethics, and they made possible the survival of the human race. Today, after the development of nuclear weapons, selfishness and aggressiveness threaten the survival of the human species.

In 1835, Alexis de Tocquiville attributed the extraordinary political and personal freedom of the people in the United States to their being intensely religious. President Franklin Rosevelt, in a famous pre-World War II speech, declared that religion was one of four essential freedoms, the others being (1) freedom from fear; (2) freedom of expression; and (3) freedom from want.

As societies evolved, religions provided a constraint on personal and societal evil. They could raise people to great heights and keep them from crashing into depression and dispair.

Today, religion in the Western World has all but disappeared. It is no longer central to European intellectual and political life. Excessive individualism, selfishness, crime, and the taking of illicit drugs have become major problems.

The idea that the world is divided into good and evil is not new. In the 3rd century, a Persian prophet, Mani, taught that *matter* is intrinsically evil, whereas *mind* is intrinsically good. St. Augustine, who was at first a Manichean, later condemned Mani's teaching but accepted the idea of the struggle between good and evil. Today, some believe that humans are innately good, whereas others believe they are innately evil.

In his 1994 book *The Moral Animal*, Robert Wright wrote, "If a mutant gene expresses the strategy 'do unto others as they've done unto you' ... an expanding circle of cooperation will suffuse the population generation after generation." Friendship, affection, and trust will hold human societies together. Aggressiveness, violence, and war will disappear in the face of humankind's innate altruism. Wright does not accept the idea that genes for altruism exist in the human genome.

Some consider crime as a disease, and war a malignant disease. St. Paul said, "The good I would, I do not, and that I would not, that I do. O wretched man that I am." Abraham, Jesus, Mohammed, and Buddha created laws that could control aggressiveness and violence, and help people achieve peace and tranquility. Did these characteristics for good result in an increase the reproductive capability of those persons whose mutant genes encoded altruism?

Han Fei (280–233 BC) did not think so. He taught that morality was not due to innate moral qualities, but to external circumstances (Lee, 1995). Chinese philosopher Xunzi (300–215 BC) believed "Human nature is evil; goodness is acquired."

Confucius and Mencius believed that people were innately good, a characteristic that led to a life of benevolence. Evil resulted from attempts to satisfy selfish desires. Emotions are innate, whereas reason and intellect are acquired, primarily the result of learning, where parents teach their offspring in words. The ability to control selfishness and greed distinguishes humans from animals.

Religions were created in response to widespread societal misery that itself brings out the evil lying dormant in our genes. Living conditions in Arabia were unbearable before Mohammed taught people to cultivate within themselves a caring, generous spirit. He taught them to make pilgrimages to Mecca, to fast during the month of Ramadan, and to surrender their entire being to God. They should give alms to the poor. If they were driven solely by selfishness, they would continue to act like wild animals. If they become aware of the needs of others, they would avoid the chaos, violence, and grasping barbarism that were prevalent in Mohammed's time. If his fellow Arabs accepted his religious principles, they would be able to overcome their innate tendencies toward aggression, rage, and violence.

Followers of the Islamic religion today are peace loving, and reject terrorism. In the April 2007 issue of the *Kuwait Times*, Muna A-Fuzai wrote, "Are we expected to be on the other side of the fence due to differences in our religions? Why do we still resort to killings under the guise of these calls for jihad?"

"We were brought up as Islamic, and as brought up with fears instilled in our minds right from the cradle-age that Israelis are our enemies. We were told and learned that they usurped Palestinian land and occupied the Aqsa mosque. Hence, we should all hate the Israelis, fight and kill them. Will killing Israelis end this long historical conflict among the Israelis, Palestinians, and the entire Gulf nationals? Most Arabs do not have anything against making peace with Israel …We should stop the murderous thoughts and hatred brewing in our minds against them, by which we gain nothing, but years wasted for no reason at all, not to mention the millions of lives lost in vain …We find no great difficulty in watching the execution of an innocent man in public and prefer to keep our ears covered, our eyes closed and our mouths shut because we are so scared that we will be killed."

"It is time to end this endless and useless struggle with the Israelis … We have a duty towards future generations, and, if a few insane souls think this world needs more people to kill and attack others meaninglessly, they are wrong and should not be tolerated anymore."

In "The Nature of the Chemical Bond," Linus Pauling wrote, "We may ask what will be the next step in the search for an understanding of the nature of life. I think it will be the elucidation of the nature of the electromagnetic phenomena involved in mental activity in relation to the molecular structure of brain tissue. I believe that thinking, both conscious and unconscious, and short-term memory, involve electromagnetic phenomena in the brain, interacting with the molecular (material) patterns of long-term memory, obtained from inheritance or experience."

Baruch Spinoza (1632–1677) believed that a unique, single substance called "God" or "nature" comprises all of reality and that its attributes account for every feature of the universe. Einstein showed that mass and energy are two forms of the same thing Spinoza referred to "energy", which cannot be created or destroyed. When Einstein was asked if he believed in God, he replied, "I believe in Spinoza's God." Plato taught that every human being is a prisoner in the "cave of civil society," in which morality is the principal constraining force, keeping aggressiveness, selfishness, and violence under control.

Immanuel Kant (1724–1804) proposed that God, freedom, and thoughts of immortality are the basis of morality. The "categorical imperative" of Kant is that morality and consciousness of one's duties and obligations should determine one's actions. It is the imperative of universal morality, which is based on freedom and reason. An individual should act the way all moral human beings should act. It is reflected in the Christian ideal that "What you should not want done to yourself, do not do unto others"; or the Hindu belief that "No man do to another that which would be repugnant to himself"; or the Jewish belief that "Thou shalt love thy neighbor as myself"; or the Zoroastrian belief that "What I hold good for myself, I should for all." The foundation of society on morality is universal.

Kant said that, "When tempted to tell a lie, one should resist the temptation, because lying is not acceptable conduct. A person should not be motivated by personal gain, fear of punishment, ridicule, or loss of reputation. He should be free and not a slave of nature, other people, or human institutions. He is free when he can control his desires and act morally. A moral person is one who must never break the law. Morality is the way to achieve happiness, and makes tolerable the fear of death and pain."

Jean-Jacques Rousseau believed that fear of death can be diminished by having a "positive" way of life, accepting pain as inevitable, and not being overwhelmed by thoughts of the inevitability of death.

Morality is the foundation of society in both Eastern and Western philosophy. Confucius (551–479 BC) believed that a moral system was the road to social harmony. Without it, no matter how hard government worked, there would still be disorder and violence:

"If you lead the people with political force and restrict them with law and punishment, they can just avoid law violation, but will have no sense of honor and shame. If you lead them with morality … they will do good of their own accord … Self-realization is a step toward peace" (Wu, 1995).

Many young people in the United States in the 1960s had views similar to those of Lao Tzu in the sixth century BC. Morality should not be the principal force in society. People rebel against authority. They should act naturally, without motives of gain or loss, praise or blame, reward, or punishment. Rules of conduct would only result in conflict, repression, and rebellion. People should return to the state in which they were born and remain *uncarved blocks*.

Confucius advocated simplicity and frugality. Austerity reduced consumerism and greed. Mencius (371–289 BC) promoted Confucianism in Chinese culture. He taught that love of all people, righteousness, courteousness, and wisdom are innate. His concept of duty was similar to the *categorical moral imperative* of Immanuel Kant. He rejected the idea of Daoism that there is virtue in selfishness.

Allan Bloom, in his book *Giants and Dwarfs* (Simon and Shuster, 1990), wrote that, "Hobbes and Locke brought men together as passengers on a ship whose interests are private but who all equally have the desire to keep the ship afloat. The natural man and the citizen are at opposite poles. A choice must be made between natural satisfaction and moral action, between the private and the public, between the particular and universal."

Society needs rules and will survive only if most men in society obey those rules. Mankind has a haunting fear of death. Government's reason for existence is to insure the safety of the people. In his book, Leviathan, Hobbes wrote: "Fear of things invisible is the natural seed of that which everyone in himself calleth religion."

Beginning in the seventeenth century, after the religious wars of the prior century, secularism began to evolve from religion, from Christianity in particular. Mark Lilla (2007) argued "A Great Separation took place, severing Western political philosophy decisively from cosmology and theology. It remains the most distinctive feature of the modern West to this day." Hobbes had planted the seed that political institutions need not be grounded on divine revelation. The new political thinking would concentrate on trying to keep people from harming one another. Peace was what mattered most. Hobbes created modern political philosophy. His ideas made democracy possible, characterized by the rule of law, the constitutional defense of human rights, and religious freedom. John Locke (1632–1704) promoted Hobbes' ideas.

Jean-Jacques Rousseau (1712–1778) shared Hobbes' ideas but postulated "an inner light" that, in modern terms, encodes our sense of obligations to others, our philanthropy, and every other altruistic behavior. George Wilhelm Friedrich Hegel (1770–1831) conceived of morality as a "categorical imperative": "Act only according to that maxim whereby you can at the same time will that it should become a universal law."

In The Law Of God, Remi Brague (2007) believes that religious and secular ideas should be combined in the ordering of society. This has been a challenge of Judaism, Christianity and Islam over the past 3,000 years. Christians borrowed the idea of natural law from the Greeks, citing Thomas Aquinas who considered the laws revealed by God to be "a gift to rational and free human beings ... Law is the way we act when in full possession of our freedom." For Aquinas, there is no zero-sum game between God and man, between revelation and reason, between faith and politics.

Mark Lilla believes that Christian political theology inevitably leads to tyranny and despotism. He ignores the idea that the wars of religion in the Middle Ages were not primarily theological, but related to political factors, economic interests, and the pain associated with the emergence of nation-states. From the late nineteenth century to the present time, Christian doctrine has argued for pluralism, religious freedom, civility, tolerance and the rule of law. The First Amendment of the constitution respects religious freedom and commitment to religious tolerance. The overwhelming majority of Americans accept the separation of religious authority from political authority, while respecting the fact that many politicians have principles based on their religion. They defend the commitment to religious freedom, civility in public life, and tolerance of those with whom they disagree.

Since the time of Darwin, many have believed that moral and social instincts are innate and are modified by life experiences. The struggle is to embrace societal advances without abandoning morality. For example, humans are motivated to give aid to their fellow man. Their desire for praise or blame is acquired. Sympathy is innate and is strengthened by exercise and habit. The moral nature of man evolved with his intellect and then was strongly influenced by habit, example, instruction, and reflection. Virtuous tendencies are inherited and reinforced by cultural and societal praise or blame.

Nicholas Wade of The New York Times (March 20, 2007) cites biologists who argue that some behavior of chimpanzees are precursors of human morality. He quotes Marc Hauser, an evolutionary biologist from Harvard, who proposed in his book Moral Minds that the brain has a genetically programmed mechanism for acquiring moral rules, a universal moral grammar similar to the neural machinery for learning language. He and primatologist Francis de Waal of Emory University believe that the roots of morality can be seen in the social behavior of monkeys and apes. He argues that human morality would be impossible without certain emotional building blocks that are at work in chimp and monkey societies. Social living requires empathy, which is needed to bring aggressive actions to an end. Four kinds of behavior—empathy, the ability to follow social rules, reciprocity and peacemaking—are the basis of society.

Religion arose thousands of years after the fundamentals of morality evolved in animals. F. B. states, "The profound irony is that our noblest achievement – morality – has evolutionary ties to our most base behavior – warfare." Immanuel Kant believed that morality is based on reason, whereas David Hume argued that moral judgments proceed from emotions. DeWaal believes that, "Human behavior derives above all from fast, automated, emotional judgments, and only secondarily from slower conscious processes. Emotions are our compass, shaped by evolution." Natural selection favors organisms that survive and reproduce.

Morality can be related to brain chemistry (Laurence Tancredi, 2005). The question is whether morality has been shaped primarily by genetics, or by culture, which takes the interests of one's entire community into account. The pragmatist would answer "both."

Many people today have faith in the cultural and political progress that Christianity brought to the world. It gave birth to the values of individuality, moral universalism, reason and progress. The two World Wars brought barbarity, slaughter of millions of people, and destruction to cities of the world, and they were a major blow to the religious foundation of politics and society. God's hand no longer seemed to be guiding society. Mark Lilla called this the *Great Separation*, i.e., the separation of political legitimacy from divine revelation.

In the United States, theology has never played a major role in government. In the Moslem world, however, many still believe that God has revealed laws governing all of human life. For many Muslims, politics is based on divine revelation, which governs the status of women, parents' rights over their children, speech inciting violence, and standards of dress. Westerners have taken for granted that science, technology, urbanization, and education have led most people to abandon their traditional faith. This has not been the case in many parts of the world.

Karl Marx viewed religion as "the opium of the people," a response to economic injustice. Moral problems were defined as societal problems. Religion was exploited by oppressors to make people tolerate being poor. The twentieth century was the bloodiest of all time, and is believed by many to have resulted from the decline of religious beliefs.

Today, many Americans immerse themselves in the Bible as they face the enormous challenges of modern life. They believe that the disappearance of religion has led to the weakening of social bonds and will result in a dark future. Mark Lilla proposes that, "We have chosen to limit our politics to protecting individuals from the worst harm that they can inflict on one another, to securing fundamental liberties and providing for their basic welfare, while leaving their spiritual destinies in their own hands."

Karen Armstrong, in her book *The Spiral Staircase* (2005), wrote that "all of the world's faiths put suffering at the top of their agendas because it is an inescapable fact of human life … the religious quest is not about discovering the truth or the meaning of life, but about living as intensely as possible here and now." The concept of God and the birth of religions, she believes, are "projections of the human mind," and that Jews, Christians, Buddhists, Hindus and Moslems all produced the same kind of God, with Buddha, Mohammed and Jesus being "icons of a fulfilled humanity."

Common to all religions is empathy for the suffering of humans, overcoming innate tendencies to selfishness and self-indulgence. The Golden Rule is a universal religious belief, born of despair, horror, and vulnerability of humans. Mohammed faced the "ghastly conditions of seventh century Arabia." The fundamental religious principles of peace, forgiveness, and reconciliation have evolved to counteract the innate human tendency to violence.

According to Karen Amstrong, religions are "… a disciplined attempt to go beyond the ego … Theologians in all the great faiths have devised all kinds of myths to show that kenosis, or self-emptying, is found in the life of God. … This is the way that human nature seems to work. We are most creative and sense other possibilities that transcend our ordinary experience when we leave ourselves behind. There may be a biological reason for this. The need to protect ourselves and survive has been so strongly implanted in us by millennia of evolution that, if we deliberately flout this instinct, we enter another state of consciousness … the history of religion shows that when people develop the kind of lifestyle that restrains greed and selfishness, they experience a transcendence that has been interpreted in different ways. It has sometimes been regarded as a supernatural reality, sometimes as a personality, sometimes as wholly impersonal, and sometimes a dimension that is entirely natural to humanity."

17

Brain Dysfunction in Children

Down syndrome is a genetic disease of the brain. In normal persons, all 23 pairs of chromosomes divide in meiosis, the process by which one diploid eukaryotic cell divides to generate four haploid cells, called gametes. In 1959, Jerome Lejeune and Patricia Jacobs discovered that some patients had three rather than the normal two copies of chromosome number 21. This abnormality is called trisomy.

When one pair of chromosomes does not divide, the resulting cells have 24 chromosomes and others have 22. A fertilized egg will develop into a child with Down syndrome, where the child inherits genes on 46 chromosomes: 23 coming from the mother and 23 from the father. The extra chromosome causes physical and cognitive abnormalities, ranging from mild-to-severe learning disabilities. The symptoms are similar to those of patients with Alzheimer's disease.

The genetic abnormalities in Down syndrome involve dominant genes. Hundreds of abnormal human traits are related to dominant inheritance, whereas an equal number are recessive. An example of the latter is sickle cell anemia, a disease in which an abnormal recessive gene is inherited from each parent. If one parent does not have the recessive gene, the child will carry the sickle-cell trait, but not symptoms and signs of the disease.

Patients with Alzheimer's disease and Down syndrome have extra Alzheimer precursor proteins (APPs) in their brain as a result of mutations in the genes encoding APP. The APP gene is located on chromosome 21. Six different mutations have been identified, usually dominant. Most early onset Alzheimer's disease is caused by mutations in the presenilin 1 and presenilin 2 genes.

Cleavage of APPs by protease enzymes, called secretases, produces peptide fragments called amyloid-β (Aβ). Aβ forms dense extracellular aggregates, called amyloid plaques, that cause neuronal death. APPs can be found in the human brain without causing deleterious effects. APPs protect brain cells from injury. Normally, the enzyme α-secretase cleaves APPs so that amyloid-β peptide is not produced. Amyloid β peptide is toxic to brain cells.

Down syndrome occurs in 1 per 800–1,000 births, but the risk in women >35 years of age is >1 in 250. All women regardless of age should have their fetus screened for the presence of the abnormal genes. Genetic testing is most accurate between the 16th week and the 18th week of pregnancy. The test is usually performed in the first trimester to enable parents more time to make choices about the course of action if there is a positive test. There is a serum marker, called pregnancy associated plasma protein-A (PAPP-A) that has a detection rate of approximately 95% of all cases of Down syndrome, with a false-positive rate of 5%.

In studies with F-18 deoxyglucose (FDG)/positron emission tomography (PET), increased neuronal activity was observed in the temporal lobe in patients with Down syndrome (Haier et al., 2003). Global neuronal activity in the brain was not significantly different in these patients from that in a control group. Six regions of the brain had higher FDG uptake in the patients with Down syndrome than in the control group.

17.1. Autism

Another common childhood disease affecting mental function is autism, originally called Kanner's syndrome, named after a Johns Hopkins pediatrician who defined the disease. The diagnosis is based on examination of the child's behavior and neurological examination. Recent reviews estimate a prevalence of 1–2 per 1,000 births for autism and close to 6 per 1,000 for autism spectrum disorder (ASD).

Many body systems are involved: *neurological, gastrointestinal, endocrine, immune, and developmental.* The number of children diagnosed as autistic has increased dramatically since the 1980s; the increase is due to better diagnosis, wider public awareness of the disease, and changes in diagnostic criteria.

A child with autism may seem to develop normally and then withdraw and become indifferent to interactions with other people, and focus intently on one thing to the exclusion of others for long periods. A child with autism may fail to respond to his name and avoid eye contact with other people. He or she does not respond normally to a person's tone of voice or facial expressions and seems to totally lack empathy.

Autistic children often engage in repetitive movements such as rocking and twirling, or in self-abusive behavior such as biting or head banging. They start speaking later than other children and may refer to themselves by name instead of "I" or "me." They do not know how to play with other children, and may talk about a narrow range of favorite topics, with little regard for the person to whom they are speaking.

Many children with autism have a reduced sensitivity to pain, and are abnormally sensitive to sound, touch, or other sensory stimulation. They resist being cuddled or hugged. Many are mentally retarded. The disease is usually diagnosed before the age of three; and it is a life-long, although treatable, disease.

They have difficulty making friends, and limited nonverbal communication. Instead of identifying people on the basis of their faces, persons with autism use only the lower face, the mouth, and other specific portions of the face to identify others. The cerebellum of some children with autism can be either hypo- or hyperplastic. There may be evidence of dysfunction of the prefrontal cortex and its connections to the parietal lobe.

They have problems of perception of spatial relationships, which suggests involvement of the parietal lobes. Identification of objects, faces, and gestures involves the temporal lobes. "Eye gaze is an important conditioning and teaching tool for young children, but autistic children tend to avoid the direct gaze. From a very early age, they do not orient toward a face" (Susan Bookheimer, UCLA). When normal children looked at faces, functional magnetic resonance imaging (fMRI) showed increased activity in the ventrolateral prefrontal cortex, whereas autistic children did not. Different brain regions are involved in different social activities, and abnormalities of these systems occur in patients with autism. Both genetic and environmental factors are involved.

Researchers at the Kennedy Krieger Institute in Baltimore, MD, found that children with autism have increased white matter in the motor region of the brain. In normally developing children, increased white matter is correlated with improved motor skills, but later may contribute to motor dysfunction, as well as abnormal socialization and communication.

Children with autism often experience difficulties with motor control and learning complex motor skills, such as riding a tricycle; pumping their legs on a swing; or buttoning, zipping, and tying shoe laces. They may do well in academic areas, such as math.

The serotonergic neuronal system may play a role in autism, possibly related to an increased frequency and preferential transmission of the long allele of the serotonin transporter gene. Vanderbilt University researchers report the discovery of 19 mutations in the serotonin transporter gene (SLC6A4) in 120 families having more than one boy affected with autism (Hsien Lei et al.). These mutations are rare, affecting only a small group of individuals with autism.

The identical twin of a child with autism has a 65% chance of having autism and about a 90% chance of having one of the ASDs, such as Asperger's syndrome, according to Dr. Eric Hollander, Director of the Seaver Autism Center at Mount Sinai School of Medicine in New York. Siblings of autistic children, especially male siblings (boys are 4 times more likely than girls to have autism), have a higher risk of autism.

Anomalies in the amygdala also have been found. fMRI reveals involvement of these regions in autism. Some with Asperger syndrome, a variant of autism, have abnormal interconnections between limbic and paralimbic regions, the cerebellum, and the visual cortices, revealed when they are asked to identify emotions by facial expressions.

Patients with autism have decreased serotonin, dopamine, nicotine, and other neurotransmitters in certain regions of the brain, but no single abnormality. Chugani (2004) examined the evidence for anomalous development of the serotonin system in the brain of children with autism. Serotonin levels in the blood are often elevated in children with autism. Serotonin synthesis is decreased in the frontal region and thalamus on one side of the brain and increased in the dentate nucleus of the opposite cerebellum.

α-^{11}C-Methyl-L-tryptophan (AMT) is used to measure the synthesis of serotonin. Deceased accumulation of AMT in the left cerebral cortex was found in children with autism who have language disorders (Chandana, 2005). They studied 117 children with autism and found that children with left cortical AMT decreases showed severe language impairments. Children with right cortical AMT decreases showed more left and mixed handedness. The finding that there is decreased availability of serotonin receptors in some persons with autism has led to the treatment with drugs, such as 5-hydroxytryptamine 2A receptor antagonists, that increase brain serotonin. Autistic disorders are heterogeneous with multiple, mostly unknown causes.

The neurohypophyseal hormones oxytocin and vasopressin are involved in autism. Elevated levels of blood serotonin occur in one third to two thirds of persons with autism. Parents of autistic children often have elevated blood serotonin levels, as well as an increased incidence of depression, obsessions, and compulsions.

PET studies have decreased serotonin synthesis in the left frontal cortex and the left thalamus and increased serotonin synthesis in the dentate nucleus of the right cerebellum.

These studies of serotonin synthesis in autism suggest that serotonin metabolism differs according to gender and intelligence.

17.2. Rett's Disease

Rett's disease has many causes and manifestations. Increased and decreased dopamine-receptor binding has been found some patients with this disorder. Studies of the dopamine transporter with the ligand β-carbomethoxy-3 β-4-fluorophenyl tropane in healthy control subjects revealed a reduction in the binding potential of the tracer in the caudate and the putamen in patients with Rett's disease. Brasic, Wong, and Eroglu have carried out PET studies of patients with autistic disorders, including Rett's disease. The male-to-female ratio of such patients is 3–4:1. Related disorders include Asperger disease, childhood disintegrative disorder, and other developmental disorders.

Dysfunction in dopaminergic neurotransmission in the caudate and putamen is related to the abnormal movements and postures. PET studies of Rett patients with ^{11}C-methylspiperone showed low or normal values of receptor density of postsynaptic D2-like dopamine receptors in the caudate of persons with Rett disorder. Other researchers using single photon emission computed tomography (SPECT) after administration of [^{123}I]iodolisuride found increased binding of the radiotracer to the D2-dopamine receptors, indicating an increase in receptor availability.

17.3. Tourette Syndrome

Tourette syndrome is another variant of autism, named after a French physician Georges Gilles de la Tourette, who first described the disease in 1885. Tourette syndrome usually begins in childhood, more often in males than females. The frequent of tics in patients with this disease become worse with increased stress.

Eight percent of children with autism have Tourette syndrome (Baron-Cohen et al., 1999). Patients with Tourette syndrome have dysfunction of the dopaminergic and serotonergic systems. The latter may be involved in their failure to develop normal social interactions in infancy and childhood. They may become normal later.

Although molecular imaging with PET has detected abnormalities in regional cerebral metabolism and some parts of the neurotransmission process, they do not provide a comprehensive picture of brain manifestations in autism. For example, SPECT imaging with technetium Tc-99m hexamethylpropyleneamine oxide did not reveal specific abnormalities in patients with autism, although many different abnormalities were found.

18

BrainDysfunctionInAdults

Over a century ago in Amagansett, NY, George Huntington described patients with symptoms of a disease that was later named after him. As the young patients became adults, they began to have uncontrolled bodily movements, lost their intellectual faculties, and developed emotional problems. In the United States today, 250,000 people are at risk of developing Huntington's disease (HD), and 30,000 have symptoms of the disease.

More than 1,600 human diseases are linked to single gene abnormalities. In 1983, Rudolph Tanzi and James Gusella used genetic linkage analysis to detect genetic abnormalities in persons at risk of developing HD. In the fourth chromosome, the sequence cytosine–adenine–guanine (C–A–G) of nucleotides is repeated 35 times in normal people. People who are at risk of developing Huntington's disease have more than 35 repeats of this C–A–G sequence.

No one knows why this genetic abnormality leads to the death of neurons, eventually leading to jerky bodily movements and twitches, and an inability to walk, swallow, think, and talk. The greater the number of times that the C–A–G sequence is repeated, the earlier the person will develop symptoms, and the faster the disease will progress.

Each child of a parent who suffers from HD has a 50/50 chance of inheriting the genetic abnormality. If a child does not inherit the gene, he or she cannot pass it to subsequent generations, and will not develop the disease. The decision of whether to take the genetic test is very difficult. About half of the people with a parent who has HD decide not to take the genetic test, in part because there is no known treatment of HD.

In summary, HD results from genetically programmed degeneration of neurons in the basal ganglia, particularly those in the caudate nuclei and the pallidum. The cortex also is affected. The diagnosis is made by a genetic test on a blood sample. The genetic test analyzes DNA for the HD mutation by counting the number of repeats in the HD gene region.

18.1. Depression

The word *depression* can refer to a symptom or a disease. Pharmaceutical companies have been a major factor in bringing about the transformation of a symptom into a disease. Branding symptoms as diseases and unceasing advertising on television can produce enormous profits.

What in the past might have been considered normal responses to stress are now described as diseases that require medication. Shyness and depression are examples.

Some societies associate shyness with politeness, good manners, and refinement, whereas others promote the idea that shyness and depression are deplorably neglected diseases that requires drug treatment. Perceived needs are manufactured along with products to treat them. Often, the disease is manifest only by subjective abnormalities, and epidemiological studies are based solely only on subjective evidence. Large pharmaceutical companies, called "Big Pharma," spend $25 billion every year on marketing, and they employ more lobbyists in Washington than there are legislators.

The goal of research is to obtain objective, measurable molecular manifestations of undesirable traits, such as extreme aggressiveness, rage, and violence, so that diseases can be defined by molecular manifestations. Hippocrates classified depression as a disease of the brain that required medical treatment.

Today, in the United States, doctors write 123 million prescriptions every year to treat depression. More than 40 million people all over the world take antidepressant drugs. According to the National Institute of Mental Health, the number of women diagnosed as suffering from severe depression in the United States doubled from 1970 to 1992. The World Health Organization predicts that depression will become the most common disabling disorder in the world by 2020, second only to heart disease.

Tranquilizers, such as Miltown and Valium, were taken to improve personality, but because of side effects, these drugs fell into disfavor. In the 1990s, selective serotonin (5-HT) reuptake inhibitors (SSRIs), including Prozac, Zoloft, Paxil, Luvox, Celexa, and Effexor, were advertised and promoted as being able to enhance alertness and feelings of well-being. In his book *Let Them Eat Prozac*, David Healy wrote that drug firms were pushing what he believed was a myth in which depression was stated to be the result of low levels of 5-HT in the brain.

Drevets et al. (1999) found decreased binding of the positron emitting radiotracer [^{11}C] WAY-100635, to the 5-HT 1A receptor in the mesiotemporal cortex and raphe in patients with recurrent familial depression. The brains of depressed suicide victims have abnormalities of 5-HT 1A receptors, 5-HT transporter binding, and 5-HT transporter mRNA. An increase in 5-HT 1A receptors was found in the midbrain of suicide victims (Drevets et al., 1998).

Effexor is an inhibitor of the reuptake of both norepinephrine and 5-HT. Michael Thase of the University of Pittsburg found that Efforex caused a 45% remission rate in patients with depression compared with the SSRI rate of 35% and placebo of 25%. Subsequent studies showed less striking results, and hypertension was an occasional side effect.

The "bible" for defining mental illness is the publication by the American Psychiatric Association's Diagnostic and Statistical Manual of Mental Disorders (DSM). To be approved for third party funding of treatment, a malady must be listed in DSM, which was introduced in 1952. The third edition in 1980 listed 112 more disorders than the second edition in 1968. The fourth edition in 1994 added 350 more, some of which were identified only by symptoms, such a feeling bad, worrying, bearing grudges, and smoking.

The idea that depression is the result of 5-HT deficiency remains controversial. Molecular imaging is able to test this hypothesis. An analogy to measurement of 5-HT activity is the measurement of radioiodine uptake by the thyroid gland or elevated serum levels of thyroxine. These measurements help define hyper- and hypothyroidism.

According to Alan Horwitz and Jerome Wakefield, it would be extremely helpful if we could use more objective criteria to characterize mental illnesses. The extraordinarily high rates of untreated mental illness reported by community studies are false. They are based on responses to questions about symptoms, and they are not able to differentiate between reactions to normal life stress (i.e., a death, a romantic break up, work or school stress) and clinical mental illness. People often seek treatment and decide for themselves ҉҉҉҉҉ ҉҉҉҉ ҉҉҉҉҉҉҉҉҉ ҉҉҉҉҉ NORMAL ҉҉҉҉҉҉҉ ҉҉ ҉҉҉҉҉ events. Pharmaceutical companies encourage sales of antidepressants.

Many seriously depressed persons have problems with cognition. The hippocampus is 10% smaller than in other depressed patients. "Depression is associated with shrinkage of the part of the brain related to memory. People in late middle age who have experienced a major depression even once in the previous 10 years are twice as likely to develop problems in concentration, memory, or problem-solving ability after the age of 65, according to several large, epidemiological studies" (Wang, 2007).

Paroxetine (Paxil, Seroxat, Pexeva) is an SSRI that increases synaptic 5-HT levels in the treatment of depression. Approved by the Food and Drug Administration (FDA) in 1992, by 2006 there were nearly 20 million prescriptions.

When brain glucose use was measured with PET/FDG in depressed patients who had improved after treatment with paroxetine, there was increased glucose use in the dorsolateral, ventrolateral, and medial regions of the prefrontal cortex (left greater than right), parietal cortex, and dorsal anterior cingulated (Kennedy et al., 2004). Glucose use fell in the anterior and posterior insular regions on the left side, as well as in the right hippocampal and parahippocampal regions. Before treatment, the glucose metabolic activity in the prefrontal regions was lower than normal. The reduction in anterior cingulate metabolic activity represented a decrease from previously elevated metabolic levels. The investigators concluded that dysfunction in the cortical-limbic neuronal circuitry of depressed patients was reversed by successful paroxetine treatment.

Unlike dopamine, 5-HT is released into the general neuronal regions, not just into synapses, and diffuses over a much larger area to activate neuronal 5-HT receptors. 3,4-Methylenedioxy-N-methylamphetamine (MDMA), cocaine, tricyclic antidepressants, and SSRIs all inhibit the removal of 5-HT from neuronal sites.

MDMA results in a flooding of the brain with 5-HT, as well as the release of synaptic norepinephrine and dopamine, and brings about feelings of well being, comfort, tactile sensitivity, and friendly emotions.

SSRIs maintain 5-HT levels in synapses longer than normal, increasing the action of synaptic 5-HT. Enzymes called monoamine oxidase (MAO) inhibitors prevent the breakdown of 5-HT and other monoamine neurotransmitters. Although small increases are pleasant, large increases in serotonergic activity can produce a variety of symptoms: confusion, agitation, headache, coma, shivering, sweating, fever, hypertension tachycardia, nausea, diarrhea, twitching, and tremor.

Depressed persons who commit suicide have been found to have low levels of 5-HT) and 5-hydroxyindoleacetic acid (5-HIAA) in their brain stem (Mann et al., 1990). Low levels of 5-HIAA also have been found in the cerebrospinal fluid of suicide victims. This has been attributed to a decreased release of 5-HT, rather than to lower numbers 5-HT receptors. Depressed persons also have low brain levels of norepinephrine, as well as 5-HT.

The amino acid L-tryptophan cannot be synthesized in the human body, but it is an essential precursor of a number of neurotransmitters, including 5-HT. Dietary tryptophan has an effect on mood and sleeping patterns. When the 5-HT precursors

L-5-hydroxytryptophan (5-HTP) and L-tryptophan are administered to experimental animals, they develop tonic and repetitive movements, tremor, rigidity, hind limb abduction, "wet dog" shaking, head weaving from side to side, and treading motions of their forepaws. These effects are the result of 5-HT increasing motor neuron excitability.

In 1987, the FDA approved the SRRI fluoxetine (Prozac) for treatment of depression, obsessive-compulsive disorder, and bulimia nervosa. Within 2 years of its approval, 65,000 Prozac prescriptions were written every month in the United States. It had become the most widely prescribed drug in the country. Other SSRIs are sertraline (Zoloft) and paroxetine (Paxil).

Persons taking Prozac become less depressed a few weeks after starting treatment. The improvement lasts up to 38 weeks. Some persons who are not clinically depressed take Prozac for a "quick fix" to help them feel better.

In his book *Listening to Prozac*, Peter Kramer describes the spectacular rise and fall in the number of persons receiving the drug: "Prozac enjoyed the career of a true celebrity—renown, followed by rumors, then notoriety, scandal, and lawsuits, and finally a quiet rehabilitation."

Few drugs are more controversial. More adverse reactions to the drug have been reported to the FDA than any other drug in the 24-year history of the FDA's adverse drug reaction reporting system. Thomas G. Whittle and Richard Wieland (*The Story Behind Prozac …. the Killer Drug*) write, "Based on documents recently obtained by under the Freedom of Information Act, as of September 16, 1993, 28,623 reports of adverse reactions to Prozac had been received by the FDA."

In August 2004, the FDA required that all antidepressant drugs must include in their label that side effects include anxiety, agitation, panic attacks, insomnia, irritability, hostility, aggressiveness, impulsivity, akathisia (psychomotor restlessness), hypomania, and mania in adults and children. Other possible side effects include delirium, hallucinations, convulsions, violent hostility, aggression, and psychosis. Up until that time, there had been 1,885 suicide attempts and 1,734 deaths, with 1,089 of them being suicide. The FDA requires a "black box warning" (its most serious) for Prozac and other SSRIs, citing the increased risk of suicide in children and adolescents taking the drug.

Since the late 1990s, Eli Lilly and Company, manufacturer of Prozac, has been the target of innumerable lawsuits involving hundreds of millions of dollars. A metaanalysis sponsored by Eli Lilly examined the results of the administration of Prozac to 3,067 patients, and concluded, "Data from these trials do not show that fluoxetine is associated with an increased risk of suicidal acts or emergence of substantial suicidal thoughts among depressed patients." They argue that the patients would not have been treated with Prozac unless they were depressed and possibly suicidal. Depression, not Prozac, is what brought about the suicides.

They cite an example of an 8-year-old boy who mutilated himself by cutting his feet with a razor blade and tying a tie around his neck, an event that was attributed by lawyers to Prozac. The plaintiff lawyers claimed that there were >20,000 suicides that occurred in persons taking Prozac above the number who would have committed suicide if they had been left untreated. The defendant lawyers claim that of the 20 million persons taking Prozac during the 1990s, 500 would have committed suicide even if they had never taken Prozac.

In summary, measurement of 5-HT, dopamine, and epinephrine receptor availability and reuptake sites on presynaptic neurons in the brain by PET or with simpler, nonimaging, probe devices provide objective measurements that can be used in diagnosis and treatment. The same tests could be used in the diagnosis of bipolar disorder.

18.2. Bipolar Disorder

Depressed patients often suffer from bipolar disorder, which in the past was called manic-depressive illness. Patients have wide mood swings with manic alternating with depressive episodes. At times, they are extremely happy, talk fast, exhibit a high level of energy, have a reduced need for sleep, have flighty ideas, express grandiosity, elation, have poor judgment, and suffer from excessive aggressiveness and often hostility.

When depressed, they have an overwhelming feeling of emptiness or sadness, a lack of energy, a loss of interest, trouble concentrating, sleep poorly, lose their appetite, and think about dying or suicide.

In 1817, the Swede Johann Arfvedson discovered the element lithium. Its biological effect is on intracellular influx of sodium during the process of axonal depolarization, which interferes with the synthesis and reuptake of neurotransmitters. In the 1950s, it began to be used in the treatment of bipolar disorder, because it dampens neurotransmission. It enhances the reuptake of dopamine, norepinephrine, and 5-HT into neuronal vesicles, reducing their action. It also reduces release of norepinephrine from synaptic vesicles and inhibits production of cAMP. It decreases the neuronal activity excited by 5-HT, dopamine, and epinephrine.

In 1970, the FDA approved lithium carbonate. When given to a patient, episode symptoms may disappear within 1–3 weeks. Molecular imaging can be used to assess the effects of lithium.

18.3. Alzheimer's Disease (AD)

A typical patient, a 75-year-old retired professor living in a senior living center, began to have difficulty finding his room when returned from meals. He kept asking: "What's my room number?" Later, he couldn't remember names, or what people had said to him 2 minutes before. He began to forget mealtimes, became less active and at times seemed agitated and depressed. He lost all self-esteem and confidence. Despite his increasing memory loss, he maintained a good spirit, but eventually began to have bad days. He could no longer to carry out activities that he enjoyed so much in the past, such as reading and listening to concerts. After several years, he lost control of his body functions, and died soon afterward.

Alois Alzheimer presented a key paper to the meeting of the South West German Society of Alienists on the 3rd November 1906. He described a patient, Frau Auguste D, whom he had taken care of in 1901. She was a 51-year-old woman who had entered a mental hospital in Frankfurt, Germany. She was unable to answer simple questions, had difficultly with her memory, was often delirious, and had hallucinations. When she died 5 years later, an autopsy revealed the presence of sticky plaques and tangles of neurons throughout her brain. We still do not know whether the plaques are a cause or a result of the disease. Regardless, they are manifestations of the disease.

AD (senile dementia of the Alzheimer type; SDAT), the most common cause of dementia, affects 4.5 million Americans. SDAT occurs in one of five people between the ages of
75 and 84 in the United States; 42% of those >85 years of age. People with SDAT survive
only half as long as those without dementia.

The disease costs nearly $150 billion per year in the United States, counting medical
and nursing home costs. The number of affected persons is expected to increase to 13.2
million by 2050. Pharmaceutical companies spend billions of dollars trying to develop
drugs to treat SDAT. Some believe that the disease is a looming "epidemic."

In AD, there are profound disturbances in mental activities involving the medial temporal lobe. Decreased neuronal activity is seen in these regions by molecular imaging.
Buckner and colleagues (2005), using magnetic resonance imaging (MRI), measured the
volume of two regions of the brain, previously linked with age-associated changes: the
corpus callosum and the medial temporal lobe cerebral cortex.

Older adults without dementia, and individuals with mild dementia of the Alzheimer
type, were different from patients with AD. There were differences between normal aging
and Alzheimer's disease. The corpus callosum was smaller in older adults, regardless of
whether they had dementia or not. Volume reductions in the hippocampus were markedly greater in people with AD.

Executive abilities involving the cognitive processes needed to complete complex,
goal-oriented tasks decline with normal aging. These problems are associated with age-
related changes in the frontal-striatal neuronal networks, and with impaired connections
across the corpus callosum.

The severe memory loss of AD is associated with decreased neuronal activity in the
medial temporal lobe memory network, which includes the hippocampus. The decline
in long-term memory is most typical of Alzheimer's disease. Elderly persons with no
symptoms of dementia may have difficulty concentrating on one thing when distractions
are present. They become confused in complex, novel situations (Buckner, 2004).

In MRI studies of healthy subjects between the ages of 20–89, Kenji Ishii of the Tokyo
Metropolitan Institute of Gerontology found atrophy around the Sylvian fissure, inter-
hemispheric space, and the parieto-occipital sulcus, reflecting dilatation of the major
groove of the brain. Glucose use declined with age in the cerebral cortex, although the
activity in areas involved with language were preserved. The findings were consistent
with Buckner's concept that executive or performance intelligence begins to decrease at
a relatively young age, but verbal intelligence is maintained even after 60.

Pre- and postsynaptic nigrostriatal neuronal markers decreased by 8% per decade in
the case of presynaptic neurons, and a decrease of 4% per decade for postsynaptic dopaminergic neurons. The increase in regional cerebral blood flow during the performance
of verbal and tactile tasks was less in the aged compared with younger persons.

The chemical activity in the neurons and billions of synapses in the brain and elsewhere in the body consumes 20% of the body's energy needs. When a person learns
something new, specific synapses are strengthened relative to others, and consume more
energy.

In FDG studies of glucose utilization by the brain during the visual identification of
words, there is an age-related slowing of the response to visual stimulation, indicated by
a decrease in regional glucose use. The retrieval of semantic/lexical information affected
glucose use the same way in both younger and older adults. In both groups, lexical-related

activation occurs in the inferior prefrontal and occipito-temporal regions of the left hemisphere. Differential activation at different ages was observed in the left occipito-temporal pathway. The older adults had higher levels of neural activity in the striatal cortex during a visual search, and in the inferior temporal cortex during lexical decision-making. The prefrontal activation was similar for the two age groups.

The decrease in the size of the brain with age in normal persons is the result of a decrease in synapses, not neurons. High stress and hypertension contribute to intellectual decline, again related to loss of synapses. Memories are believed to be the result of long-term potentiation (LTP) of neurons. It declines with age. With aging, there is a decrease in the number of dendrites (Coleman and Flood, 1987). Synaptic deterioration with aging causes the loss of memory, but neuronal death is the more likely cause of AD. Older people who continue to be mentally active are less likely to suffer from memory loss. Neocortical loss of synapses in patients with AD is correlated with the degree of cognitive impairment (DeKosky and Scheff, 1990; Terry et al., 1995).

In 1993, when asked how close we are to an effective treatment for AD, Tanzi answered, "I wouldn't be surprised if 5 years from now we have a pretty effective drug that can slow the disease down enough so that it will be preventable in those at risk, and significantly slow down the deterioration of people who already have it."

Two other genes in familial AD are presenilin 1 and 2. This form of AD affects about 10% of patients. The APOE4 gene is linked to the more widespread, late-onset disease. What the Tanzi group has done more recently is identify another gene, like APOE4, that is associated with an increased risk in developing AD.

Working with a group of 437 families, each with two first-degree relatives affected by Alzheimer's, his group looked at genes on chromosome 9, and they identified genes producing a protein called ubiquilin. "The reason we specifically looked at ubiquilin was because it binds to and interacts with the early onset gene, presenilin 2," Tanzi explained. Variants in a ubiquilin-linked gene, UBQLN1, were associated with late-onset Alzheimer's in affected families.

In the past, it was thought that neurons could not replicate. We now know that this is not true. Neuron-forming stem cells can reproduce in the hippocampus, which is involved in memory and learning, and often affected in older people as well as those with AD. In older people and persons with AD "neural stem cells are present but not dividing, so they are not making new neurons," said Ashok K. Shetty, Ph.D., Professor of Neurosurgery at Duke and the Durham Veterans Affairs Medical Center. "We hope that by making more neurons, we can improve learning and memory" in these patients. A drug under study is Memantine, which is thought to protect the brain from the effects of high levels of glutamine. This drug may prevent the formation of tangles.

It has been known for several years that APP may play an important role in patients with AD. Some think that plaque formation triggers the disease and that tangles are a secondary effect. This is known as the amyloid-cascade hypothesis.

APP is the precursor of the amyloid-β protein that causes the typical "plaques" in the brains of patients. The protein (APP) is metabolized by two enzymes known as β- and γ-secretase. One of the products of their action is the production of amyloid β-peptide 42, which causes these molecules to stick to one another. If you stop the plaques from forming, or get rid of them once they have formed, it may arrest the disease process.

Dale Shenk, Ph.D., at Elan Pharmaceuticals in San Francisco reported that the initial results of vaccination were very promising (*Psychiatric News*, August 18, 2000). Unfortunately, he stopped the trial because of central nervous system inflammation in some of the subjects. There is still hope that a β-amyloid vaccine might eventually be a safe and effective treatment for AD, and many are pursuing research in this direction. Some peptides that bind to amyloid plaques are not toxic to human neurons in tissue culture. Other investigators are developing compounds that inhibit the enzymes involved in the production of β-amyloid, namely, β- and γ-secretase.

Another approach is to develop secretase inhibitors. Aspirin and Ibuprofen act on γ-secretase. They inhibit cleavage of amyloid precursor protein. The normal function of APP was not known. Maarten Leyssen and his colleagues think that APP stimulates the development of neuronal pathways that are essential for the proper functioning of the brain. Neuronal connections can be damaged by trauma. APP helps stimulate the development of new nerve paths.

The fruit fly is an ideal model for studying the brain, and research indicates that APP increases after brain damage in areas where new nerve paths are being formed. APP increases the production of plaques. Patients with major brain damage have increased risk of developing AD later. Patients who have had brain trauma often develop plaques that resemble those in Alzheimer patients.

William Klunk and Chester Mathis vadiolabelled a fluorescent dye, related to Congo Red, labeled with fluorine-18, that is bound by β-amyloid protein in plaques found in the brain of patients with AD. After years of research, they developed a dye with 34 times the affinity for amyloid than Congo Red. Some think that β-amyloid is the cause of AD, whereas others think it is the result. Patients with Down syndrome carry an extra copy of the amyloid gene, and they develop brain plaques and symptoms of AD when they reach middle age.

In 2002, a new tracer, called Pittsburgh Compound B (PIB), was found to clearly bind to amyloid plaques in the brains of Alzheimer's patients. Today, investigators in the United States, Europe, Asia, and Australia are studying whether new drugs can inhibit the buildup of amyloid plaque and slow the course of the disease.

Dr. Bengt Langstrom at Uppsala University in Sweden said recently, "We have succeeded in showing the presence of this substance (amyloid) in the living brain. This opens up the possibility to differentiate patients with Alzheimer's from patients with other diseases and thus offers the possibility of early treatment. We can now follow up the effect of new drugs that might have an impact on the harmful substance amyloid."

β-Amyloid protein in cerebrospinal fluid is also used as a biomarker for AD. Research in mouse models of the disease has shown that greater amounts of amyloid-containing plaques in the brain are associated with lower levels of a specific protein fragment, β-amyloid 42 (Aβ42), in cerebrospinal fluid.

Anne M. Fagan, Ph.D., and colleagues from the Washington University School of Medicine and the University of Pittsburgh, have reported the same inverse relationship between Aβ42 in the brain and in cerebrospinal fluid (CSF) (Fagan, 2006).

PET imaging with the radiotracer PIB, which binds to amyloid plaques in the brain was performed, together with samples of CSF. In patients with AD, there was avid binding of PIB in several regions, including the prefrontal cortex, precuneus, and temporal cortex.

The cerebellar cortex and brain stem white matter showed little specific binding, a pattern consistent with the initial PIB study in AD brain, according to the study.

Three of the seven positive PIB scans (with corresponding low values of Aβ42) had no symptoms of dementia. Fagan said that it is possible that these subjects have pre-clinical AD. If they develop AD, it would mean that we may have a way to screen individuals for the presence of AD pathology before the appearance of clinical symptoms."

Too much β-amyloid is the result of the action of other enzymes with other roles, so that blocking them may cause side effects. Using cultured human and mouse cells, as well as test tube assays, University of Texas Southwestern researchers singled out how just one portion of the enzyme, a protein called nicastrin, is involved in the pathway that produces β-amyloid, thereby leading to AD. They hope next to work on ways to specifically block nicastrin (August 12 issue of *Cell*).

The four drugs now on the market for treating AD include Aricept by Eisai and Pfizer, Exelon by Novartis, Razadyne by Johnson & Johnson, and Namenda by Forest Laboratories. These drugs inhibit the enzyme cholinesterase that is responsible for the breakdown of acetylcholine in the brain. These drugs seem to postpone the worsening of Alzheimer's symptoms for 6–12 months in about half of the people who take them.

Although these drugs have limited effectiveness, their sales were about $1.4 million per year in the Untied States alone in 2006. Todd Golde of the Mayo Clinic has said, "It's scary if you look at the trials that got these drugs approved. The change in mental status was so small. The average caregiver of a patient would have no way of knowing whether there was any difference." The changes can be as minor as a better ability to dress oneself or take out the trash. With the high failure rate in developing drugs, it is unlikely that many of the drugs under development will make it to widespread clinical use.

The drug Dlurizan, a product of Myriad Genetics, was developed in one of 22 research projects supported by the National Institutes of Health (NIH) in Bethesda, MD, and it has an anti-inflammatory effect. Dlurizan is thought to lower the production of plaques.

The "amyloid cascade" describes the process by which the brain develops plaque, which contains a protein called amyloid, which occurs as a breakdown product of a substance called APP.

In an article in the September 30, 2004, issue of *Neuron*, Randy L. Buckner points out that what he calls *executive function* commonly decreases with normal aging. Wisdom and experience persist, or may even increase with age. Memory loss is thought to be the result of impairment of frontal-striatal neuronal networks, together with deterioration of the corpus callosum that connects the cerebral hemispheres. The accelerated memory loss of AD is also associated with deterioration of the medial temporal lobe, including the hippocampus.

In summary, MRI and computed tomography (CT) can provide information about the shape, position, or volume of various brain structures. Structural techniques include MRI and CT. PET and fMRI can reveal the neuronal activity in regions of the brain that are usually affected. The brains of people with Alzheimer's shrink significantly as the disease progresses. Shrinkage of the brain and decreased neuronal activity measured by F-18 FDG occurs early in the course of AD. These studies also can also be used to monitor the progression of the disease.

18.4. Parkinson's Disease (PD)

The symptoms and signs of AD and PD often occur in the same patient. The similarity of many of the manifestations of PD and AD are so great that they seem to be variants of the same pathological processes.

In PD, there is a loss of neurons in the substantia nigra and elsewhere in the brain in association with the presence of protein deposits in the cytoplasm of neurons (Lewy bodies) and thread-like proteinaceous inclusions within Lewy neurites.

For years, it was thought that environmental factors, possibly toxins, were the cause of PD. In 1997, a gene associated with the disease was identified. Mutations in certain genes (the "parkin" or "synuclein" gene) can result from environmental toxins, such as exposure to pesticides. The LRRK2 gene was isolated by Andrew Singleton, Ph.D., of the National Institute on Aging at the NIH. A mutation of this gene is found in 5% of patients with inherited PD. Tatiana Foroud of the Indiana University School of Medicine examined 767 patients with PD from 358 families. The LRRK2 gene, located on chromosome 12, is one of five genes in patients with PD that are abnormal.

There is a decrease in neuromelanin in the substantia nigra. There is proliferation of microglia in regions of dopaminergic terminal loss in patients with untreated PD. Microglia are activated in response to the inflammatory process in the brain, and they are believed to play a role in the production of neuroprotective substances, such as nerve growth factors, in response to tissue injury and inflammation.

Melanin is a pigment in hair, skin, eyes, and feathers, and it protects tissue from damage by UV light. Neuromelanin is the dark pigment present in pigment-bearing neurons of four deep brain nuclei: the substantia nigra, the locus ceruleus, the dorsal motor nucleus of the vagus nerve (cranial nerve X), and the median raphe nucleus of the pons. These nuclei are not pigmented at the time of birth, but they develop pigmentation during maturation to adulthood.

Neuromelanin is a by-product of the synthesis of monoamine neurotransmitters. The loss of pigmented neurons is seen in a variety of neurodegenerative diseases. In PD, there is massive loss of dopamine producing pigmented neurons in the substantia nigra. In AD, there is an almost complete loss of the norepinephrine-producing pigmented neurons of the locus ceruleus.

Neurons using dopamine are concentrated in the midbrain in the substantia nigra. Kawamura and colleagues at the Tokyo Metropolitan Institute of Gerontology examined the binding of $[^{11}C]CFT$ to striatal dopamine transporters (DATs) by PET, and they found an age-dependent decrease in the striatal uptake of $[^{11}C]CFTk3$). This was related to both. In patients with PD, dopaminergic neurons degenerate, resulting in an absence of dopamine in the basal ganglia, especially the putamen.

Patients with PD do not move their arms when they walk, and they are slow in performing all their activities. Eighty percent of the patients have a tremor at rest that disappears when they move. There is a general stiffening of the body. When properly diagnosed, treatment is begun with L-dopa that increases dopamine in the brain, often with a dramatic improvement in symptoms. Its effectiveness decreases after about 4–5 years of treatment. L-dopa, the precursor of dopamine, was introduced >30 years ago. The fact that administration of the dopamine precursor, L-DOPA greatly relieves the patient's symptoms, is an early example of how chemical characterization of a disease can lead to molecular therapy.

The role of the dopaminergic system in PD was first recognized in 1961, when W. Birkmayer and O. Hornykewicz found a markedly reduced concentration of dopamine in the liver of myelin and dopaminergic nigrostriatal dopamine-containing neurons in post-mortem brains of patients. They found a direct correlation between the degree of dopamine deficiency and neuronal degeneration. This discovery led eventually to the administration of L-dopa in treatment of Parkinson's patients. Treated patients showed greater effects of L-dopa than do normal persons, suggesting that they may have an increased number of available receptors.

PD has a prevalence of approximately 0.5–1% among persons 65–69 years of age, rising to 1–3% among persons 80 years of age and older. Many patients have symptoms and signs of idiopathic Parkinsonism, as do patients with postencephalitic, drug-induced, and arteriosclerotic parkinsonism.

Five hundred thousand Americans, as well as Muhammad Ali, Michael J. Fox, Janet Reno, and Pope John Paul II were afflicted with the disease. PD needs to be distinguished from benign essential tremors, as well as from the slowness and tremor that often accompany aging. Other diseases such as Wilson's disease and cerebrovascular disease may mimic PD.

Multiple system atrophy (MSA) is similar to PD. Gerhard and colleagues in the United Kingdom found increased binding of the PET tracer [^{11}C]R-PK11195 in the dorsolateral prefrontal cortex, putamen, pallidum, pons, and substantia nigra, regions that are involved in MSA. PET imaging provides evidence of microglial activation in multiple system atrophy.

Ouchi and colleagues of the Positron Medical Center and Department of Neurosurgery, Hamamatsu Medical Center and Central Research Laboratory, Hamamatsu Photonics K.K., Japan, studied the changes in the nigrostriatal and mesocortical dopaminergic systems related to gait in patients with PD. They examined DAT availability in the striatum and extrastriatal region during walking exercise in normal subjects and age-matched unmedicated patients with PD. [^{11}C]CFT uptake by the DATs in the putamen was decreased to a greater extent in normal persons than patients with PD, whereas a significant reduction in [^{11}C]CFT uptake was not found in the putamen but in the caudate and orbitofrontal cortex in PD patients. These results are evidence that DAT availability is reduced in the nigrostriatal projection area. Activation in the medial striatum and the mesocortical dopaminergic system may be related to gait disturbances in PD patients.

Patients with PD cannot achieve the proper balance between stimulatory and inhibitory neuronal activity affecting movement. Selegiline (Eldepryl, Movergan), also known as deprenyl, is an antioxidant drug that blocks MAOB, an enzyme that degrades dopamine. Until recently, selegiline, or deprenyl, was the drug most commonly used in early onset disease and in combination with levodopa. It seems to delay the onset of the freezing of the patient's gait.

An interesting approach to the treatment of patients with PD is gene therapy, by using viruses to carry growth factors into the basal ganglia. In 1986, Rudolf Tanzi discovered the APP gene, identified as the first known Alzheimer's gene. In 1993, he isolated the gene associated with Wilson's disease, a neurological disease that results from copper toxicity. He also has collaborated on the identification of a gene associated with a familial form of Lou Gehrig's disease, or amyotrophic lateral sclerosis, a progressive neuromuscular disease that weakens and eventually destroys motor neurons that control movements, such as walking and talking.

"Gene therapy offers another mechanism in which to get growth factors into the brain," said Diane Murphy, a neurodegenerative disease specialist at the NIH. Other investigators implant genes that increase production of an enzyme that converts levodopa into dopamine.

[^{11}C]R-PK11195 binds to peripheral benzodiazepine receptors and makes possible the imaging of activated microglia. The density of dopamine neuronal terminals can be examined with the DAT tracer [^{11}C] w-βcarbomethoxy-3β-4-fluorophenyl tropane ([^{11}C] CFT). The [^{11}C] CFT binding in the midbrain on the opposite side of the brain from the affected side in patients with unilateral PD was higher than in healthy control subjects. The microglial response to inflammation is thought to contribute to the progression of the disease, and it provides a target for neuroprotective drug therapy.

About 3% of babies are born with birth defects in the United States each year, according to the U.S. Centers for Disease Control and Prevention. The goal of genetic profiling is to identify and assess the risks of an inherited disorder in every newborn child. The search continues to find the relationship between genotype and phenotype, and to learn which genes encode the phenotype. Molecular imaging can play a major role in partnering with genetic neuroscience.

In summary, PET or SPECT can detect abnormalities of the dopaminergic system in the basal ganglia, particularly the putamen.

18.5. Schizophrenia

Kevin, a 24-year-old black man, was arrested when the police arrived after his mother called 911. He was taken to San Francisco General Hospital (SFGH), talking about getting a gun, and killing people. He was sullen and uncommunicative as the police forcibly took him to the hospital. "I don't want to talk to any people."

He had been diagnosed as schizophrenic when he was 16 or 17 years old, but psychiatrist Heather Hall at SFGH doubted that the diagnosis was correct. He stared blankly into the distance, as many schizophrenic patients do. "He looked really, really sad." Hall believed the correct diagnosis was depression. His mother had called the police when a quarrel with his friend escalated into a fight. The diagnosis of mental illness is often difficult and arbitrary.

Ethnicity often affects the diagnosis. Both Ken and Heather Hall are African Americans. "Maybe because I am an African American psychiatrist, he was able to show me a little more of himself so I could make an accurate diagnosis and change his treatment .... the people who work in our unit are sensitive to the issues of African Americans. We are much more likely to look at our patients with eyes that aren't clouded by preconceived notions."

Another time, a psychiatrist making rounds with residents, asked the interpreter to check whether her Spanish-speaking patient was likely to commit suicide. The patient responded: "I feel so bad I could die." The interpreter declared, "She's suicidal."

Another psychiatrist related the story of a 30-year-old black woman who was talking fast, calling people at all hours of the day, seeming not to need sleep-classic symptoms of bipolar disorder. She had been diagnosed as schizophrenic, but the psychiatrist was dubious: "How could a woman with a college education, who is euphoric, speaks rapidly, and has a decreased need for sleep be schizophrenic?" The correct diagnosis was bipolar illness. Physicians can often have different impressions when a patient comes from a different cultural setting.

Egan, Weinberger, and colleagues at the National Institute of Mental Health reported that the gene responsible for the control of glutamate may help determine the risk for schizophrenia. Glutamate has long been thought to play a role in schizophrenia. The gene codes for the glutamate receptor GRM3 is responsible for regulating glutamate in synapses. The amount of glutamate remaining in the synapse affects cognition. Egan et al. implicate the GRM3 gene, which affects glutamate transmission, brain physiology, and cognition.

Meltzer, Park, and Kessler of Vanderbilt University School of Medicine, and Northwestern University, Chicago, have focused on the cognitive abnormalities in patients with schizophrenia. Recognition of the central importance of cognitive impairment to schizophrenia further diminished in the 1950s, after the development of antipsychotic drugs such as chlorpromazine and haloperidol, which had the ability to improve delusions and hallucinations, although, at times, causing significant extrapyramidal side effects. The diagnostic schema for schizophrenia during the 1970s and 1980s emphasized delusions and hallucinations (positive symptoms), which, in about 70% of the patients, respond to the haloperidol-like drugs. These drugs were found not to improve cognitive function and, sometimes, to impair some aspects of cognition, such as memory.

Subsequent research on the cognitive impairment of schizophrenic patients revealed defects in cognition, involving attention, executive function, secondary (storage) memory, working memory, and semantic memory. Components of the cognitive deficit were found to be present during childhood and early adolescence in a mild form. The decline in cognition was thought to develop during the period of several months to several years before onset of psychosis.

C. Robert Cloninger of the Departments of Psychiatry and Genetics, Washington University School of Medicine, St. Louis, MO, believes that specific genes influence susceptibility to schizophrenia, a disease that affects nearly 1% of people throughout the world. Chumakov and colleagues also have found genes associated with increased susceptibility to schizophrenia. A gene, "G72," located on 13q34, interacts with the gene for D-amino acid oxidase on 12q24 to regulate glutaminergic signaling through the N-methyl-D-aspartate (NMDA) receptor pathway. The gene dysbindin on 6p22.3 (3) and the gene neuregulin 1 on 8p also increase susceptibility to schizophrenia and may operate via the same NMDA mechanism.

Twin studies indicate that susceptibility to schizophrenia is strongly heritable, even if children are reared apart from their biological parents. If one twin has schizophrenia, the risk of schizophrenia in the co-twin is greater in monozygotic twins (45%) than in dizygotic twins (15%). Forty percent of the monozygotic co-twins of a person with schizophrenia are clinically normal. The inheritance pattern of schizophrenia suggested that multiple genes interact with one another and with environmental factors to influence susceptibility. This has been confirmed by >20 genome-wide linkage scans in more than 1,200 families of schizophrenics.

In genetic studies, researchers are taking a closer look at RNA, which has a greater significance than was previously thought. RNA's role was only thought to be carrying genetic information from DNA in the nucleus to the places in the cell where proteins are made. RNA selects the amino acids from which proteins are synthesized. Transfer RNA carries the information from DNA to provide building materials for the encoded proteins. There is now increasing evidence that genes are also RNA factories. The number of different RNA's may be as high as 37,000, instructing the production of proteins (Encyclopedia of DNA, called ENCODE). What are called microRNAs regulate the activity of at least a third of human protein-encoding genes. Thus, the whole genetic and evolutionary story involves more than DNA.

One of the most active areas of research in molecular imaging is the study of the effect of age. Suhara and colleagues used PET to study the age-related changes in human muscarinic acetylcholine receptors. In persons between the ages of 18 and 75, uptake of $[^{11}C]N$-methyl piperidyl benzilate continuously increased in all brain areas with the exception of the cerebellum. The binding of the tracer to acetylcholine receptors in eight brain regions (pons, hippocampus, frontal cortex, striatum, temporal cortex, thalamus, occipital cortex, and parietal cortex) showed an age-related decrease of about 45%.

Perhaps the most important effect to date of advances in molecular imaging is in the design, development and assessment of the value of PET and SPECT studies An important role is in selecting patients for these trials. Symptoms alone often provide inadequate criteria for diagnosis. Patients with schizophrenia must be separated from those patients suffering from depression or other mental dysfunction.

Zeber and colleagues, in the October 2004 issue of the journal *Social Psychiatry and Psychiatric Epidemiology*, pointed out the problems. Darrel Regier, American Psychiatric Association, also agrees that cultural differences between patients and doctors can result in misdiagnosis.

"I believe bias exists, and there is a risk that a psychiatrist with a different cultural background from a patient can misinterpret the expression of psychiatric symptoms," he said. "If you have a very religious group of patients and a very secular psychiatrist who thinks beliefs in spirits or hearing the voice of God is not normal, you are going to have errors."

"If you have an African American suffering from paranoia, it may be considered a healthy "paranoia" based the patient's perception of society", said Zeber, who works at the Veterans Affairs Department's Health Services Research and Development Center in San Antonio. Misdiagnosis can be due to a different cultural perspective of bizarre behavior. Failure to appreciate that social and cultural factors play a role in the diagnostic process can lead to some patients being erroneously stereotyped, which causes errors in treatment.

Many schizophrenic patients are depressed, and may experience bipolar-like symptoms. When mood disorders are a major feature of the illness, it is called schizoaffective disorder, with elements of schizophrenia and mood disorders in the same individual.

The diagnosis of schizophrenia is subjective, based on three types of symptoms: (1) positive, (2) disorganized, and (3) negative. Positive symptoms include delusions and hallucinations. Patients believe that people are reading their thoughts or plotting against them, secretly monitoring their actions, and threatening them. They believe that others are controlling their minds. Disorganized symptoms include confused thinking, speaking, and inappropriate behavior. They have problems speaking coherent sentences or ⬛⬛⬛⬛⬛⬛⬛⬛⬛ ⬛⬛⬛⬛⬛⬛⬛⬛ ⬛⬛⬛ ⬛⬛⬛⬛⬛ ⬛⬛⬛⬛⬛⬛⬛⬛⬛⬛ ⬛⬛ ⬛⬛⬛⬛⬛⬛⬛ ⬛⬛⬛⬛⬛⬛⬛ ⬛⬛ ⬛⬛ ⬛⬛⬛⬛⬛⬛⬛ ⬛⬛ circles. Negative symptoms include lack of emotions or their poor expression. They have difficulty in starting or following through on activities. There is little pleasure or interest in life. They are not able to reason correctly, remember things, or plan how to achieve their goals. Attention and motivation are lacking.

Some patients with psychotic and affective disorders, but not severely demented, are misdiagnosed as having Alzheimer's disease (SDAT). Paul Eugen Bleuler, a Swiss psychiatrist, recognized that they did not have SDAT. He introduced the concept of schizophrenia as primarily neurocognitive disorder. The categorization of patients as being schizophrenic was strengthened in the 1950s, when antipsychotic drugs, chlorpromazine and haloperidol, were found to decrease delusions and hallucinations (Herbert Y. Meltzer, Vanderbilt University and Northwestern University in Chicago).

During the 1970s and 1980s, the positive symptoms—delusions and hallucinations—were emphasized. Seventy percent of schizophrenic patients respond to antipsychotic drugs, such as haloperidol. These drugs did not to improve cognitive function, such as memory, or motor functions, but only the positive symptoms. The difficulties with attention, executive function, long-term memory, working memory, and semantic memory were unaffected. These deficits vary widely among different persons with schizophrenia; about 15% of schizophrenic patients have cognitive impairment.

Schizophrenia is a huge societal challenge, in addition to causing great suffering to the patients and their families. The patients usually require lifetime care, at a great cost to society. The cost of caring for schizophrenic patients in the United States is $20 billion a year. The first signs of the disease often occur in the teenage years or early twenties. The World Health Organization has identified schizophrenia as one of the ten most debilitating diseases affecting humans. In the United States, 2.2 million American adults, or 1.1% of the population age who are 18 and older, are schizophrenic.

Schizophrenia occurs all over the world. "Not many people realize that there are over 15 million sufferers of schizophrenia in China alone, with 60% of them too incapacitated to work, and another 20% needing constant hospitalization," said Hannah Hong Xue. She recently led a team in discovering a new pattern of DNA sequence variations among individuals with predispositions to various diseases, including a gene strongly linked with schizophrenia. (*Molecular Psychiatry*, December, 2003). "Our discovery of the gene, one of five to cause schizophrenia so far identified, opens new paths for the effective treatment of the disease … As we start to improve our understanding of schizophrenia, customized and hopefully affordable drugs can be tailored to the patient's exact medical needs É prevention may become a possibility."

For centuries, people have used herbal medicines to help patients with schizophrenia. In India and the East Indies, Rauwolfia serpentina offered some hope, and was used for a long time to treat Chandra, i.e., "moon disease or lunacy." The roots of Rauwolfia yielded the drug reserpine, the first major tranquilizer, adopted by Western medicine in 1943 and first used to treat hypertensive patients in the 1950s. Before that, it had been used in India in millions of patients.

In 1954, Dr. Robert Wilkins was the first Western physician to report his results in treating hypertensive patients. The same year, Nathan Kline described the anti-psychotic effects of reserpine, and it soon replaced electric shock and lobotomy in the treatment of schizophrenia. Many synthetic alkaloids were developed as tranquilizers. Reserpine is a dopamine antagonist, that is, it inhibits the actions of dopamine by binding to post-synaptic dopamine receptors. The cell bodies of neurons that produce dopamine are in a midbrain region, the substantia nigra. Dopamine antagonists "turn down" the dopamine activity in the basal ganglia and other regions of the brain.

Reserpine blocks the uptake (and storage) of norepinephrine and dopamine into synaptic vesicles by inhibiting the vesicular monoamine transporters. Brodie and colleagues found that reserpine also depleted serotonin from body tissues, including the brain. Their studies included patients with carcinoid tumors that produce serotonin, after there are metastases to the liver.

One of the complications of reserpine treatment was the appearance of symptoms and signs similar to patients with PD. This finding directed attention to dopamine in PD.

Reserpine profoundly reduces synaptic dopamine. Normally, unbound dopamine in the synapse is broken down by the enzyme MAO, or taken back into presynaptic neurons. Once taken back, it is sequestered inside presynaptic vesicles, where it is no longer exposed to MAO. Reserpine blocks this reuptake of dopamine and other neurotransmitters, and it remains in synapses, where it is broken down by MAO.

Amphetamine blocks the uptake of dopamine, 5-HT, and norepinephrine by presynaptic transporters, increasing the extracellular levels of these monoamines. D-amphetamine acts primarily on the dopaminergic systems, whereas L-amphetamine acts primarily on norepinephrinergic neurons. The stimulant effects of D-amphetamine on behavior are linked to enhanced dopaminergic activity, primarily in the mesolimbic system of neurons.

Amphetamine, sold as *Benzedrine* for recreational purposes, was inhaled to produce a euphoric stimulating effect. Others abused the drug by cracking open the inhalers and swallowing the paper strips covering the Benzedrine. "Bennies" are still used by military pilots to fight fatigue and increase alertness. In 1959, after decades of abuse by the civilian population, the FDA banned the sale of benzedrine inhalers and limited the use of amphetamine to physician prescriptions. Crystallized methamphetamine (*ice*) was also used in World War II to help keep soldiers and pilots awake. In patients with schizophrenia, large doses made their symptoms worse. The negative symptoms in patients with schizophrenia—social withdrawal, apathy, and anhedonia—are associated with low dopamine levels in certain regons of the brain. Manic subjects have increased levels.

Increased dopaminergic activity in the brain is believed to be important in the pathogenesis of schizophrenia, although the label schizophrenia is applied to a heterogeneous group of patients with differing pathologies. The dopamine hypothesis states that abnormalities of the dopaminergic system are common in schizophrenia. Post-mortem studies have shown increased numbers of dopamine receptors in the brains of schizophrenic patients. Autopsy studies often reveal enlarged ventricles and a thinner cerebral cortex.

PET scanning makes it possible to measure the availability of dopamine receptors in the patients with schizophrenia compared with nonschizophrenics. The striatum, limbic system, and the cortex are the principal locations of dopamine neurons. Studies of young, drug-naive schizophrenic patients failed to demonstrate abnormalities in the densities or affinities of D2 dopamine receptors in the basal ganglia, contradicting the earlier theory that elevated densities of D2 dopamine receptors are a major pathophysiological mechanism in schizophrenia. The prefrontal cortex of schizophrenic patients has decreased levels of D1, D3, and D4 receptors.

D̲1̲ ̲d̲o̲p̲a̲m̲i̲n̲e̲ ̲receptors have been correlated with the severity of negative symptoms. Although current research seems to indicate that dopamine is involved in schizophrenia, it is difficult to determine its exact role. Increased levels of dopamine in the striatum, as well as the prefrontal cortex, are responsible for the positive symptoms. The correlation between D1 receptors and negative symptoms is explained by the fact that the prefrontal cortex sends neuronal impulses to the rest of the brain. D1 receptors may be involved in the production of movement in response to signals initiated in the cortex. A decrease in these receptors would result in an inhibiting effect on behavior that would be similar to the negative symptoms. Both the positive and negative emotional symptoms may be related to the activity of dopamine in the limbic system, which has a high concentration of D3 and D4 receptors. Patients with schizophrenia have smaller frontal lobes, larger ventricles, and a higher incidence of head injury during childhood. (*The Dopamine Hypothesis of Schizophrenia*. Anne Frederickson).

The dopamine hypothesis is based chiefly on the observation that neuroleptic drugs that block dopamine receptors on post-synaptic dopaminergic neurons reduce the symptoms of schizophrenic patients. Schizophrenia has been characterized by abnormally low prefrontal dopamine activity (causing deficit symptoms), which leads to excessive dopamine activity in mesolimbic dopamine neurons (causing positive symptoms).

Dopamine is involved in feelings of pleasure. Increased dopaminergic activity is not only involved in schizophrenia but also in attention deficit hyperactivity disorder and other psychiatric conditions. The serotonergic system also is thought to play a role.

"Neurotransmitters relay messages to the brain at two speeds: fast and slow. Our new findings indicate that brain receptors that respond to dopamine have two slow modes: one mode that takes place over a period of minutes and a second—newly discovered— that lasts for hours. This effect continues for as long as dopamine remains in the system" (Jean-Martin Beaulieu, 2005).

All addictive drugs, including cocaine and amphetamines, raise dopamine levels. Dopamine plays a major role in the biological effects of all abused drugs that induce highs and lead to addiction. They raise dopamine levels in the nucleus accumbens, the reward center of the brain, activated in anticipation of rewards, such as food, and during feelings of well-being. Dopamine relays messages to the brain within minutes but affect the brain over long periods.

A regulatory protein, β-arrestin 2, is involved in desensitization of receptor signals, and dopamine-related behavior. Prolonged stimulation of D2 receptors leads to inactivation of a regulatory protein called Akt. "Our results provide direct physiologically relevant evidence for the emerging concept that beta-arrestin 2 not only controls desensitization but also participates in slow synaptic transmission by acting as a scaffold for signaling molecules in response to dopamine receptor activation" (Marc G. Caron, Duke University).

"Our results provide a pathway by which dopamine receptor activation leads to the expression of dopamine-associated behaviors." Akt plays an important role in other processes in the body, including inflammation, cell death, and cell proliferation. Similar mechanisms underlie other diseases.

The dopamine hypothesis in schizophrenia is supported by three findings: (1) All effective neuroleptic drugs are D2-dopamine receptor antagonists; (2) functional abnormalities of mental functions related to dopamine alteration are found in schizophrenic patients; and (3) drugs that increase dopaminergic activity induce psychosis in normal individuals or exacerbate symptoms in schizophrenic patients.

Other molecular abnormalities are seen in schizophrenia: (1) Serotonergic neurotransmission measured by the concentration of 5-HT and its metabolite 5-HIAA and the density of 5-HT1A or 5-HT2A receptors in post-mortem brain specimens; (2) hormonal and behavioral responses to serotonergic drugs; and (3) abnormalities of 5-HT and its metabolites in blood or CSF.

Schizophrenic patients had a significantly greater reduction in amphetamine-induced striatal radiotracer binding of the radioligand [^{123}I]iodobenzamide, imaged by SPECT, than did normal control subjects, interpreted as showing increased dopamine secretion in response to amphetamine in the schizophrenic patients.

Taminga and colleagues (1992) identified abnormalities of FDG accumulation in limbic circuits of the brain of psychotic schizophrenic patients compared with normal controls. The patients had a significantly lower regional metabolism of glucose in the hippocampus and the anterior cingulate cortex, but not in the neocortical or extrapyramidal system.

McGowan and colleagues at Hammersmith Hospital in London found increased [^{18}F] fluorodopa accumulation in the ventral striatum of the basal ganglia of patients with schizophrenia, consistent with increased dopamine release (McGowan et al., 2004). There is increased dopamine synthesis in the striatum and medial prefrontal cortex in schizophrenic patients, supporting the hypothesis of dopaminergic dysregulation.

Molecular imaging with PET and SPECT of postsynaptic dopamine and serotonin receptors and presynaptic transporters help psychiatrists and psychologists go beyond subjective assessment of symptoms, signs, inappropriate thoughts, emotions, and behavior, but they have not yet begun to apply PET and SPECT in the every day clinical care of patients. They do use molecular imaging to differentiate mild cognitive impairment from AD.

Some believe that advances in biological psychiatry have been exaggerated; yet, no one would deny the enormous advances that have occurred in psychopharmacology. McHugh and Slavney (1984) have cautioned, "Mental experiences are not things in the brain, but rather events that relate to the brain and to many other things besides."

Using PET, Farde et al. found that D2 receptor occupancy in schizophrenic patients was reduced by 20–67% of the pretreatment levels in those successfully treated with clozapine. It was reduced to 80–90% of original values that occurred with other typical neuroleptics. He concluded that D2 receptor antagonism alone does not explain the efficacy of clozapine therapy. At low doses (125–172 mg/day), clozapine occupied more than 80% of 5-HT 2A receptors.

Studies of the density of specific serotonin receptors in the brain suggested that serotonergic dysfunction might play a role in schizophrenia, but clinical trails of serotonergic precursors, agonists, or nonspecific antagonists were ineffective. Nevertheless, it was thought that some of the beneficial effects of clozapine in improving schizophrenic symptoms could be attributed to its ability to block serotonin receptors.

5-HT effects cognition, memory, perception, attention, mood, aggression, sexual drive, appetite; energy level, pain sensitivity, endocrine function, and sleep. The effects of drugs on these functions, many of which are abnormal in schizophrenia, led to the "serotonin-dopamine" hypothesis. Schizophrenia might be related to enhanced dopaminergic and serotonergic neurotransmission in subcortical areas.

In studies of 5-HT receptor density in post-mortem brain tissues from patients with schizophrenia, Bennett et al. found a decrease in [³H]lysergic acid diethylamide (LSD) binding in Brodmann areas 6, 8, 11, 44, and 47. LSD produces hallucinations by disrupting the action of the neurotransmitter serotonin, although precisely how it does this is unclear. LSD acts on certain groups of serotonin receptors, especially in the cerebral cortex and locus ceruleus.

Whitaker et al. found increased [³H]LSD binding in unmedicated schizophrenics in cortical regions 4, 10, and 11. [³H]LSD binds to six 5-HT receptors with high affinity, which makes it difficult to determine which type of receptor is responsible for the drug's effects. Using the 5-HT antagonist [³H]ketanserin as a ligand, Mita et al. found a decrease in binding in cortical area 9. [³H]Ketanserin has high affinity for both $\alpha2$-adrenergic receptors and tetrabenazine-sensitive sites.

Arora and Meltzer and Laruelle et al., using [³H]spiperone, found a decrease in binding in cortical areas 8 and 9. Joyce et al., using quantitative receptor autoradiography, reported an increase of ^{125}I-LSD and [³H]ketanserin-labeled "5-HT2" receptors in temporal and posterior cingulate, frontal, and parietal cortices, the ventral putamen, nucleus accumbens and hippocampus, but not in the caudate nucleus or motor, prefrontal, entorhinal, or anterior cingulate cortices.

Serotonergic neurons modulate dopaminergic activity and vice versa in the ventral tegmentum, substantia nigra, and medial and dorsal raphe. Multiple types of 5-HT and dopamaine receptors may be involved. The 5-HT1A, 5-HT2A, and D2 receptors are the most important in regulating effect of 5-HT on the dopaminergic system.

One can conclude that schizophrenic patients are a diverse population with many clinical and brain chemical differences among them. They should be characterized as individuals by measuring regional brain chemistry, rather than by putting them in a diagnostic *box* called schizophrenia. Cognitive dysfunction, disorganization, negative and positive symptoms, motor abnormalities, decreased sexual development, insomnia, and compulsive behaviors differ from one patient to another.

In the United States, Medicare has spent more money on antipsychotic drugs than on any other class of pharmaceuticals—including antibiotics, AIDS drugs, or medications for hypertension (Lagnado, 2007). One reason is that these drugs are given to elderly patients without symptoms of AD and other forms of dementia. The off-label use of antipsychotic drugs are reimbursed by Medicare (Centers for Medicare & Medicaid Services [CMS]) whether or not the people are psychotic. CMS stated that 21% of nursing home patients who do not have psychosis are treated with antipsychotic drugs. In New York City, the figure is closer to 70%. The $122 billion nursing home industry's use of drugs includes both typical and atypical antipsychotic drugs, the latter having sales of $11.7 billion in 2006.

The American Health Care Association, which represents for-profit, investor-owned, and nonprofit nursing homes, says that facilities must "work closely with doctors to ensure that medications prescribed are meeting the individual needs of each patient."

Now it is possible to measure the brain chemistry of mental disorders and put treatment on a more scientific basis.

18.6. Catatonia

In November 1950, I studied the effect of bulbocapnine on rats, and published the results in the *Archives of Neurology and Psychiatry*. Bulbocapnine produces all of the motor manifestations of catatonic schizophrenia in human beings. The drug is an aporphine isoquinoline alkaloid that decreases dopaminergic activity in the brain, and inhibits tyrosine hydroxylase, the rate-limiting enzyme in the catecholamine biosynthesis.

Catatonia was first described by Karl Kahlbaum in 1874 and was subsequently classified as a manifestation of schizophrenia by Eugene Bleuler in 1911. It is manifest by statuesque posturing, muscular immobility, mutism, and what seems to be stupor. The muscles are held in a pliant state called waxy flexibility, and the catatonic person obediently permits himself to be rearranged into awkward positions that may subsequently be held for hours.

In 1933, Buchman and Richter (1933) showed that the experimental catatonia produced in monkeys by bulbocapnine could be terminated by the injection of cocaine. The animals behaved normally at once and did not return to the catatonic state. It is now thought that the primary mechanism of cocaine is to block the DAT, which increases the concentration of synaptic dopamine. The effect of dopamine on interneuronal signaling is increased. While taking amphetamine for several days damages dopaminergic neurons in the basal ganglia, cocaine leaves the caudate unscathed.

Our studies showed that the bulbocapnine-induced catatonic state in wild Norway rats could be temporarily interrupted by intense auditory stimulation, electric shocking that resulted in fighting among the rats, and by immersion in deep water, which necessitated their swimming to safety. Oliver Sacks showed subsequently that catatonia resulting from encephalitis in human beings could be interrupted by the administration of L-dopa, which increases dopamine production and release.

Anthony Grace of the Department of Neuroscience and Psychiatry at the University of Pittsburgh has advanced the concept that impulses from the cortex stimulate dopamine release from subcortical structures. Grace proposed that cortically stimulated dopamine release is bound by presynaptic DATs that result in a decrease in synaptic dopamine levels. Breier and colleagues (NIH) studied the effect of amphetamine in stimulating dopamine release. In primates, doubling the amphetamine dose decreased the binding of $[^{11}C]$ raclopride to postsynaptic dopamine receptors by half, as a result of occupancy by endogenously produced dopamine.

After the first imaging of dopamine receptors in the living human brain by PET in 1983, striking changes were found with aging. D2 dopamine and S2 5-HT receptors in the caudate nucleus, putamen, and the frontal cortex fell between the ages of 19–73 years with different rates for men and women. Many other neurotransmitter receptors decline with age, presumably a result of degeneration of neurons. Age-related decreases in dopamine receptor subtypes also are seen in the hippocampus and entorhinal cortex (J. Comp. Neurol .456:176–183, 2003).

Loss of neurons and pathological lesions, including plaques or tangles, or abnormal cross-links can cause cognitive aging and diseases, such as SDAT. Dean Wong of Johns Hopkins used 3-N-[^{11}C]methylspiperone to measure D2 dopamine receptors and serotonin 5-HT 2A receptors in patients with schizophrenia, Tourette syndrome, Rett syndrome, Lesch-Nyhan syndrome, and bipolar illness. Wong and colleagues also carried out the first measurements of presynaptic DATs by using [^{11}C]WIN.

Based on autopsy studies, it was thought that schizophrenia is a manifestation of increased activity of dopaminergic neurons. Recent data suggest that there are abnormalities of the dopaminergic system in patients clinically diagnosed as schizophrenic. What seems clear is that neuroleptic drugs that block dopamine receptors on postsynaptic dopaminergic neurons reduce the symptoms, possibly related to a reduction in dopamine activity in mesolimbic neurons. Post-mortem studies have shown increased availability of dopamine receptor densities in the brains of schizophrenic patients. Schizophrenia has also been characterized by abnormally low prefrontal dopamine activity (causing deficit symptoms), which results in the excessive dopamine activity in mesolimbic dopamine neurons (causing positive symptoms).

19

Conclusion

"In former days, when men sailed only by observation of the stars, they could cast along the shores of the old continent or cross a few Mediterranean seas; but before the ocean could be traversed and the new world discovered, the use of the mariner's needle as a more faithful and certain guide, had to be found out" Francis Bacon (1561–1626).

The basic premise of this book is that we can create a molecular theory of mental illness, analogous to the germ theory of disease. Molecular imaging makes it possible to search for patterns in the chemical processes in the brain that are related to violence and other forms of mental illness, involving hormones, neurotransmitters, neuroreceptors, reuptake sites, ions, peptides, and proteins.

Violent behavior often follows brain trauma, brain tumors, encephalitis, stroke, multiple sclerosis, Alzheimer's disease, Huntington's chorea, hypoglycemia, and other metabolic disorders.

Aggressiveness results from the failure of inhibitory neurochemical processes or exaggeration of stimulatory processes in brain regions, such as the orbito-frontal cortex, the septal area, hippocampus, amygdala, caudate nucleus, thalamus, ventro-medial and posterior hypothalamus, midbrain tegmentum, pons, and the fastigial nuclei and anterior lobe of the cerebellum.

Two diseases that were defined half a century ago can illustrate a molecular theory of mental disease. The rate of accumulation of radioactive iodine by the thyroid could be measured by radiation detectors pointed at the neck. Patients, who had a disease subsequently called hyperthyroidism were found to have an increased accumulation of radioactive iodine, whereas radioiodine uptake was low in patients with hypothyroidism. Patients with hyperthyroidism were nervousness and anxious, and suffered from weight loss, whereas hypothyroid patients were lethargic and had impaired mental functions. These two diseases illustrate the fundamental principle that if one can measure a chemical process in some part of the body, there will be patients in which the chemical process is abnormally slow, and others in which it is abnormally fast (Fig. 19.1).

Dopamine, serotonin, and endogenous enkephalins (opiates) modulate neuronal activity. If you drop dopamine on isolated neurons, transmission of neuronal impulses is inhibited. Dopamine and serotonin counteract the effect of the stimulatory monoamines, epinephrine and norepinephrine.

The molecular theory of mental illness is based on molecular imaging, which uses the tracer principle to create quantifiable images of the distribution of a radioactive tracer

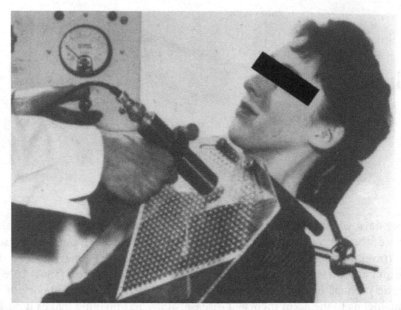

Fig. 19.1 Half a century ago, measurement of radioactive iodine in the thyroid gland with a Geiger–Mueller counter placed at different points indicated on a plastic grid over the patient's neck.

within regions of the brain. Magnetic resonance imaging (MRI), computed axial tomography, and ultrasound primarily reveal structure and function, but they are being extended to apply the tracer principle as well. MRI tracers can be activated in vivo by the action of endogenous enzymes. These so-called molecular chaperones are attached by covalent bonds to the MRI tracer. The strength of the MRI signal reflects the amount of enzyme activity at that site (Thomas Meade, Northwestern University).

In 1983, positron emission tomography was used for the first time to image dopamine receptor binding in the basal ganglia of a living human. The enthusiasm of scientists and the public that greeted this news was greater than that following the invention of electroencephalography (EEG) by Hans Berger, an Austrian psychiatrist, in 1924. Berger measured the electrical activity of the brain with electrodes attached to a person's head, which reflected the electrical action potentials traveling along billions of neuronal fibers. The EEG signals reflect the person's state of mind as information travels throughout the brain.

The use of the EEG to study the mysterious mind was met with skepticism. In 1940, Charles Sherrington wrote, "The mind is something with such manifold variety, such fleeting changes, such countless nuances, such wealth of combinations, such heights and depths of mood, such sweeps of passion, such vistas of imagination, that the submission of some electrical potentials recognizable in nerve-centers as correlative to all these may seem to the special student of the mind almost derisory." What would Sherrington say if he could see how molecular imaging was being used today to study the relationship between mind and brain?

Sherrington said, "As we contemplate the shape of psychiatry in the twenty-first century, one of the greatest risks we face is reductionism ... There is a great need ... to avoid a reductionistic approach that doesn't take the whole person into account ... everybody loves to oversimplify things ... Psychiatry must not become a house divided, with psychosocial specialists in one camp and neuroscientists in another ... Both groups need to work closely together." (Glen O. Gabbard, M.D., Baylor College of Medicine).

Psychiatrists Paul McHugh and Phillip Slavney stress how important it is to integrate the different ways of viewing mental disease, which falls into four categories: diagnosis, biology, behavior, and life stories. Molecular imaging can contribute to all four categories.

McHugh and Slavney caution, "Psychiatry may always find itself involved in battles, most often and most injuriously from partisans of one perspective who fail to give legitimacy to another. One can let a hundred flowers blossom, a hundred schools of thought contend' if the ground from which the flowers emerge is well surveyed and constantly under cultivation."

Over a century ago, William James wrote, "Our inner faculties are adapted in advance to the features of the world in which we dwell, adapted so as to secure our safety and prosperity in its midst ... Not only are our capacities for forming new habits, for remembering sequences, and for abstracting general properties from things, and associating their usual consequences with them exactly the faculties needed for steering us in this world of mixed variety and uniformity, but also our emotions and instincts are adapted to the very special features of that world. Mental action is uniformly and absolutely a function of brain-action, varying as the latter varies as effect to cause. The uniform correlation of brain-states with mind-states is a law of nature. We need to know just what sort of goings-on occur when thought corresponds to a change in the brain."

"Between a mental and physical event there will be an immediate relation, the expression of which, if we had it, would be the elementary psycho-physic law."

Molecular imaging portrays the molecules of the mind as they create anger, disgust, fear, joy, sadness, and surprise. These emotions can be correlated with measurable molecular processes in different brain regions, including the amygdala, in the fight-or-flight response. Signals arise at the amygdala from the higher centers of the brain when a person faces danger.

When a person faces danger, signals from the higher centers of the brain pump adrenaline, the hormone of stress, into the circulating blood. "There are two systems for analyzing risk: an automatic, intuitive system and a more thoughtful analysis" (Paul Slovic, University of Oregon). After a few seconds, the rational response follows the emotional response.

In 2005, Pope Benedict XVI presented a challenge to people all over the world: "A united humanity must confront the many troubling problems of the present time—from the menace of terrorism to the humiliating poverty in which millions of human beings live, from the proliferation of weapons to the pandemics and the environmental destruction, which threatens the future of our planet. The men and women in our technological age risk becoming victims of their own intellectual and technical achievements, ending up in spiritual barrenness and emptiness of heart." Science can be our guide, just as the magnetic compass guided the ancient mariners.

Violent and destructive behaviors are the result of biological, as well as cultural and political factors. In 1958, Christopher Dawson wrote, "Our generation has been forced to realize how fragile and unsubstantial are the barriers that separate civilization from the forces of destruction. We have learned that barbarism is not a picturesque myth or a half-forgotten memory of a long-passed stage of history, but an ugly underlying reality that may erupt with shattering force whenever the moral authority of a civilization loses its control" (Religion and the Rise of Western Culture, Doubleday, NY).

   "If any man there be, not content to rest in and use the knowledge which has already been discovered, aspires to penetrate further … to seek, not probable conjectures, but certain and demonstrable knowledge, I invite all such to join with me, that we may find a way into nature's inner chambers" (Bacon and Organum, 1620).

References

P. Aebersold. The cyclotron; a nuclear transformer. Radiology, 1942.

P. Aebersold, J.H. Lawrence. The physiological effects of neutron rays. Ann Rev Physiol, 1942.

F. Ajami. The Foreigner's Gift: The Americans, The Arabs, and The Iraqis in Iraq, 2007.

A. Alavi et al. First FDG/PET images of brain. Semin Nucl Med, 1981.

C. Anderson was awarded the Nobel Prize in Physics, 1932.

Annual Review of Pharmacology and Toxicology, April 2000

H. Arendt. The Life of the Mind, Harcourt Brace Jovanovich, 1971

K. Armstrong. The Spiral Staircase, Anchor Books, 2005.

A. Aron et al. Stony Brook University. Reward, motivation and emotion systems associated with early-stage intense romantic love. J Neurophysiol, 2005.

E. Aserinsky, N. Kleitman. Regularly occurring periods of eye motility, and concomitant phenomena, during sleep.S cience,1 953.

Avshalom Caspi gene variant, family factors can raise conduct disorder risk. Psychiatric News, 2004.

J. Axelrod. U.S. National Library of Medicine. Profiles in Science: The Julius Axelrod Papers.

F. Bacon. Novum Organum, 1620.

F. Bacon. The Secret of Nature, 1561–1626.

R. Bailey. Born to Be Wild? The Role of Genes in Antisocial Behavior, 2002.

S. Baron-Cohen, C. Mortimore, J. Moriarty, J. Izaguirre, M. Robertson. The prevalence of Gilles de la Tourette's syndromein c hildrena nda dolescentsw itha utism. JCh ildP sycholP sychiatry, 40, 213–218, 1999.

J.-M.B eaulieu, A β-arrestin 2 signaling complex mediates lithium action on behavior. Cell, 2005

A. Bechara, H. Damasio, A.R. Damasio. Emotion, Decision Making and the Orbitofrontal Cortex. Oxford University Press, 2000.

A.H. Becquerel. The Nobel Prize in Physics, 1903.

M. Behrmann, G. Avidan. Trends in Cognitive Sciences, 2005.

M. Bender, M.D. received the Cassen Award Society of Nuclear Medicine, 1980.

M.R. Bennett. Monoaminergic synapses and schizophrenia: 45 years of neuroleptics. J Psychopharmacol, 1998.

H. Benson. The Mind/Body Effect, 1979.

H. Benson. The Benson-Henry Institute for Mind Body Medicine, Massachusetts General Hospital.

Berman, Gladue, Taylor. The effects of hormones, Type A behavior pattern, and provocation on aggression in men. Motiv Emot, 1993.

C. Bernard. An Introduction to the Study of Experimental Medicine, 1865.

P.C. Bernhardt. Curr Dir Psychol Sci, 1997.

Bice, Wagner, Frost, et al. A simplified detection system for neuroreceptor studies in the human brain. *J Nucl Med* 2, 184–191, 1986.

A. Bloom, Giants and Dwarfs, Simon & Shuster, 1990.

N. Bohr was awarded the Nobel Prize in Physics, 1922.

T.M. Bosley et al. The visual cortex of patients recovering from ischemic lesions that caused visual field defects. Ann Neurol, 1987

D. Botstein, R.L. White, M. Skolnick, R.W. Davis, Am J Hum Genet, 1980.

R. Boyce. The Independent, UK, 2007.

R. Brague. Law of God, University of Chicago Press, 2007.

S. Brenner, winner of the Nobel Prize in 2002.

J. Bronowko. The Origins of Knowledge and Imagination. Yale University Press, 1977.

G. Brownell, Sweet. A single pair of simple radiation detectors for localization of brain tumors. Nucleonics, 11, 40, 1953.

G.L. Brownell et al. Positron tomography and nuclear magnetic resonance imaging. Science, 1982.

H.G. Brunner et al. Abnormal behavior associated with a point mutation in the structural gene for the enxyme monoamine oxidase (MAOA). Science, 1988.

R.B uckner.N ature,2 004.

R. Buckner. Cerebral Cortex, 2005.

C.D.R Burt, S.H. Snyder Dopamine receptor binding predicts clinical and pharmacological potencies of antischizophrenic drugs. Science, 1976.

M. Calvin was awarded the Nobel Prize in Chemistry, 1961 for his discovery of photosynthesis.

W.B. Cannon. The Wisdomo fth eB ody. Norton, New York, 1932.

A. Carllson, P. Greengard. Awarded the Nobel Prize in Medicine 2000 "for their discoveries concerning signal transduction in the nervous system".

M.G. Caron. A broader role for ß-Arrestins. Sciensce, STKE 2003.

M.G. Caron. G-protein-coupled receptor regulatory mechanisms. September, 2005.

A. Caspi et al. Influence of life stress on depression: moderation by a polymorphism in the 5-HTT gene. Science, 2003.

B. Cassen. The rectilinear scintillation scanner, 1950.

R.B. Cattell. Personality: a Systematic, Theoretical, and Factual Study. New York, 1950.

M. Chase. Wall Street Journal, 2008.

D. Chopak. Quantum Healing. Bantam Books, 1989.

Christopher Walsh and Geoff Woods. ASPM is a major determinant of cerebral cortical size. Nat Genetics, 2002.

P.M. Churchland. Matter and Consciousness: a Contemporary Introduction to the Philosophy of Mind. MIT Press,1 988.

C.R. Cloninger. The discovery of susceptibility genes for mental disorders. Proc Natl Acad Sci, 2002.

C.R. Cloninger. The discovery of susceptibility genes for mental disorders, 2004.

E.F. Coccaro. Serotonergic studies in patients with affective and personality disorders. Correlates with suicidal and impulsive aggressive behavior. Arch Gen Psychiatry, 1979.

E.F. Coccaro et al. Serotonergic studies in patients with affective and personality disorders. Correlates with suicidal and impulsive aggressive behavior. Arch Gen Psychiatry, 46, 1989.

J. Cohen. Supercomputing in medicine, 1997.

G.D.C ommentary,2 007.

M. Corbetta et al. Neural basis and recovery of spatial attention deficits in spatial neglect. Nat Neurosci, 2005.

Cornish, P. Kalivas. J Neurosci, 20, 2000.

A.M. Costa. Chief of the U.N. Office on Drugs.

J.T. Coyle. The glutamatergic dysfunction hypothesis of schizophrenia. Harvard Rev Psychiatry, 1996.

M. Curie. The Nobel Prize in Physics in 1903 and in Chemistry, 1911.

E. Curley. The Collected Works of Spinoza, Princeton, 1938.

C. Darwin. The Descent Man, 1871.

P. David. Hamilton: Craig Venter's genome and our brave new world. PLoS, 2007.

C. Dawson. Religion and the Rise of Western Culture. Doubleday, NY

DeCode Genetics From Genes To Drugs.

R. Degrandpre. The Cult of Pharmacology, Duke University Press, 2007.

S.T. DeKosky, S.W. Scheff, Synapse loss in frontal cortex biopsies in Alzheimer's disease: correlation with cognitive severity. Ann Neurol, 1990.

C. DeLisi. Genomes: 15 Years Later. Human Genome News, 2001.

C. DeLisi. Genomes: 15 Years Later, 2003.

D.A. Descartes' Error: Emotion, Reason and the Human Brain, 1994.

F.B.M. de Waal. Primates and Philosophers: How Morality Evolved, Princeton University Press, 2006.

K. Dewhurst. Dr Thomas Sydenham (1624–1689): his life and original writings, University of California Press, Los Angeles, 1966.

Diagnostic and Statistical Manual of Mental Disorders. American Psychiatric Association, 1994.

N. Doba. Role of the cerebellum and the vestibular apparatus in regulation of orthostatic reflexes in the cat. Circ Res, 1974.

H. Donald Burns, R.F. Dannals, B. Langström, et al. 3-N-C-11 Methyl)Spiperone, A Ligand Binding to Dopamine Receptors: Radiochemical Synthesis and Biodistribution Studies in Mice. J Nucl Med, 1984.

Drevetse ta l. JN eurosci, 18(18), 7394–7401, 998, 1998.

Drevetse ta l. BiolP sychiatry, 46(10), 1375–1387, 1999.

M.F. Egan et al. Variation in GRM3 affects cognition, prefrontal glutamate, and risk for schizophrenia. Proc Natl Acad Sci, 2004.

P. Ehrlich. Readings in Pharmacology. Little Brown, 1962.

A. Einstein, The Nobel Prize in Physics, 1921.

J. Elkes. Psychopharmacology: Finding One's Way, 1992

J. Elkes. Mind, Brain, Body, and Behavior: Foundations of Neuroscience and Behavior, 2004.

Emerson's Essays T.Y. Cowell Co.

K.I. Erickson, A.F. Kramer. Brain Cogn, 2002.

A.M.F agan. AnnN eurol,2 006.

M.J. Farah. Patient-based Approaches to Cognitive Neuroscience, 2000

L. Farde et al. D1- and D2-dopamine receptor occupancy during treatment with conventional and atypical neuroleptics. J Psychopharmacol, 1989.

N. Ferguson, The War of the World, Penguin, 2007

E. Fermi was awarded the Nobel Prize in Physics, 1990.

E. Fermi received the Nobel Prize in 1938 for "his discovery of new radioactive elements produced by neutron irradiation".

H. Fisher. Reward, motivation and emotion systems associated with early-stage intense romantic love. J Neurophysiol,2 005.

H. Fisher. Wall St J, 2007.

O.F lanagan. Consciousness Reconsidered, MIT Press, 1992.

D.G. Flood, P.D. Coleman. Neuron numbers and dendritic extent in normal aging and Alzheimer's disease. Neurobiol Aging 8, 521–545, 1987.

E. Florey. GABA: History and Perspectives. Can J Physiol Pharmacol, 1991.

D. Foley. The influence of genetic factors and life stress on depression among adolescent girls. Psychiatric News, September 3, 2004.

M. Frayn. The Human Touch. Henry Holt and Co., 2007.

A. Frederickson. The dopamine hypothesis of schizophrenia, 1998.

A. Frederickson. The dopamine hypothesis of schizophrenia, 1989.

G.O. Gabbard. The Mind-Brain Interface.

J.H.G addam,1 957

G. Gamow, Proposed Alpher-Bethe-Gamow theory of the atom as consisting of alpha, beta, and gamma radiation,1 928.

Gamma et al. Mental disorders in current and former heavy ecstasy (MDMA) users. Addiction, 2005.

H. Gardner. Leading Minds. Basic Books, 1995.

A. Garrod. The Incidence of Alkaptonuria: a Study in Chemical Individuality, 1902.

O. Gefvert et al. Different corticostriatal patterns of L-DOPA utilization in patients with untreated schizophrenia and patients treated with classical antipsychotics or clozapine. Scand J Psychol, 2003.

A. Gerhard, Neurology, 61, 686–689, 2003.

T. Golde. Alzheimer Dis Assoc Disord, 2007.

W. Gombrowicz. New York Review of Books, 2006.

Van Goozen, Frijda, Van de Poll. Anger and aggression in policewomen and administrative employees: an experimental study. Soc Sci Inf, 1995.

R.R.G riffiths et al. Mushrooms create universal 'Mystical' experience. Psychopharmacology, 2006.

J. Haidt. The Happiness Hypothesis: Finding Modern Truth in Ancient Wisdom. Basic Books, 2006.

D.A.H ammoud.R adiology,2 007. HapMapP haseI Ip ublished

L. Harris, The Suicide of Reason. Basic Books, 2007

M. Hauser. Moral Minds: How Nature Designed Our Universal Sense of Right and Wrong, 2006.

G. de Hevesy was awarded the Nobel Prize in Chemistry, 1943.

G. Hevesy, F. Paneth. The Solubility of Lead Sulfide and Lead Chromate, 1913.

T. Hobbes. The Leviathon, 1660.

G.D. Honeydagger et al. Differences in frontal cortical activation by a working memory task after substitution of risperidone for typical antipsychotic drugs in patients with schizophrenia. Proc Nat Acad Sci, 1999.

O.H ornykewicz,1 966.

A. Horwitz, J. Wakefield. The Epidemic in Mental Illness: Clinical Fact or Survey Artifact?, Winter, Contexts Magazine,2 006.

Y.-Yu. Huang et al. An Association between a Functional Polymorphism in the Monoamine Oxidase A Gene Promoter, Impulsive Traits and Early Abuse Experiences. Neuropsychopharmacology, 2004.

G. Huntington, Medical and Surgical Reporter of Philadelphia: April 13, 1872.

C. Hynes, Wall Street J, 2006.

T. Ido, J.S. Fowler, A.P. Wolf, M. Reivich, D. Kuhl. Labeled 2-deoxy-d-glucose analogs. F-18 labeled 2-deoxy-2-fluoro-d-glucose, 2-deoxy-2-fluoro-d-mannose and C-14 2-deoxy-2-fluoro-d-glucose. J Label Comp Radiopharm,1 978.

K. Ishii. Digital Signal Processing, 2007.

R.H. Jackson. United States Representative to the International Conference on Military Trials, London, 1945.

W. James. Principles of Psychology, 1890.

W. James. Pragmatism.N ewI mpression, Longmans,G reena ndC o. New York, 1914.

Jensen, V. Elwood. Received Lasker Award for Basic Medical Research, 2004.

F. Joliot, I. Joliot-Curie won the Nobel Prize in Chemistry in 1935 "in recognition of their synthesis of new radioactivee lements".

F. Joliot, I. Joliot-Curie were awarded the Nobel Prize in Physics, 1935.

C. Jones, USA Today, 2007.

D. Kahneman Heuristics and Biases: The Psychology of Intuitive Judgment. Cambridge University Press, 2002.

M.K amend iscoveredc arbon-14.

E. Kandel. A nobel laureate. The Search for Memory, 2008.

A. Karni, D. Sagi. The time course of learning a visual skill, 1993.

S.H. Kennedyeta l.T reatingD epressionE ffectively, MartinD unitz, Londona ndN ew York, 2004.

S.S. Kety The Control of the Cerebral Circulation. Circulation, 1964.

S. Kim, H.N. Wagner Jr., V.L. Villemagne, P.-F. Kao, R.F. Dannals, H.T. Ravert, T. Joh, R.B. Dixon, A.C. Civelek. J Nucl Med, 1997.

N.A. Klein. Brain Chemistry Linked to Aggressive Personality. Science Daily, 2007.

N. Kline, quoted by Merton Sandler, Monoamine Oxidase Inhibitors in Depression: History and Mythology. J Psychopharmacol,1 990.

W.E. Klunk et al. Imaging brain amyloid in Alzheimer's disease with Pittsburgh Compound-B. Ann Neurol, 2005.

S. Kosslyn. Image and Brain. MIT Press, 1994.

S.M. Kosslyn et al. The Case for Mental Imagery, 2006.

H.K osterlitz,J .H ughes

Kraepelin. Emil. Manic-Depressive Insanity and Paranoia. Classics in Psychiatry, 1921

P. Kramer. Listening to Prozac, Penguin Books, 1994.

P.D. Kramer. Listening to Prozac (Knopf), 1993; Against Depression (Viking), 2005.

M.J. Kuhar, G.K. Aghajanian. Selective Accumulation of 3H-Serotonin by Nerve Terminals of Raphe Neurones: An autoradiographic Study. Nature, 1973.

Kuhl, E. David. Nucl Med Pioneer, 1976.

D. Kuhl, R. Edwards. Tomography in nuclear medicine. Radiology, 1963.

D. Kuhl, R. Edwards. Image separation rdioisotope scanning. Radiology, 1963

L. Lagnado,Wall Street J, 2007.

B. Lahn. The Scientist, 2005.

J.N. Langley. Drugs act by binding to molecular receptors. J Physiol, 1978.

Laruelle, cited by G.J.E Schmitt. The striatal dopamine transporter in first-episode, drug-naive schizophrenic patients: evaluation by the new SPECT-ligand[99mTc]TRODAT-1. J Psychopharmacol, 2005.

J.H. Lawrence, the brother of Ernest, made the first clinical therapeutic application of an artificial radionuclide when he used phosphorus-32 to treat leukemia, 1936.

E. Lawrence was awarded the Nobel Prize in Physics, 1939.

J.H. Lawrence. The therapeutic use of artificially produced radioactive substances, radiophosphorus, radiostrontium, radioiodine, with special reference to leukemia and allied diseases. Radiology, 1942.

J.H. Lawrence, E.O. Lawrence. The biological action of neutron rays. Proc Nat Acad Sci, 1936

E. Lawrence, M.S. Livingston. The Production of High Speed Light Ions Without the Use of High Voltages. Phys Rev,1 932

T. Leary. Your Brain Is God, 2003.

W.-C. Lee. Great Thinkers of The Eastern World, Harper Collins, 1995.

R. Lemov. World as Laboratory, Hill and Wong, 2006.

D.-L. Liaoa, Neuropsychobiology, 50(4), 284–287, 2004.

I. Lieberburg. Attacking Alzheimer's disease on multiple fronts, 2001.

M. Lilla. The Stillborn God: Religion, Politics and the Modern West. Knopf, Sept, 2007.

L. Lindström. Increased dopamine synthesis rate in medial prefrontal cortex and striatum in schizophrenia indicated by (b-11C) DOPA and PET. Biol Psychiatry, 2000.

O. Loewi. Winner of the Nobel Prize in Medicine, 1936.

E. London. PET/FDG studies in addicts, 2004.

K. Lorenz. On Aggression. Bantom Books, 1966.

K.L orenz.O n Aggression,2 002.

G. MacDonough. After the Reich, Basic Books, 2007

D.J. Madden, J. Spaniol, W.L. Whiting, B. Bucur, J.M. Provenzale, R. Cabeza, L.E. White, S.A. Huettel, Adult age differences in the functional neuroanatomy of visual attention: a combined fMRI and DTI study. Neurobiol Aging,2 007.

J.J. Mann et al. Annals of the New York Academy of Sciences, 1990.

P. Maquet. Regional cerebral glucose metabolism in children with deterioration of one or more cognitive functions and continuous spike-and-wave discharges during sleep. Brain, 1995.

R. Mario, Capecchi. Nobel Lecture, 2007

C. McDonald. Qualitym easuresa nde lectronicm edicals ystems. JAMA, 282, 1181–1182, 1999.

P.L. McGeer, E. McGeer. Immunotherapy for Alzheimer's disease. Sci Aging Knowl Environ, 2004.

S. McGowan. Presynaptic dopaminergic dysfunction in schizophrenia: a positron emission tomographic [18F] fluorodopa study. Arch Gen Psychiatry, 2004.

S. McGowan et al. Presynaptic dopaminergic dysfunction in schizophrenia: a positron emission tomographic [18F]fluorodopa study. Arch Gen Psychiatry, 2004.

J.L. McGrew. The Report of the National Commission on Marihuana and Drug Abuse, 2004.

P. McHugh, P. Slavney. The Perspectives of Psychiatry, Johns Hopkins Press, 1984.

B. McNaughton, M. Wilson. Hunting for Meaning After Midnight. Science, 2007

T. Meade et al. Cell permeability of MR contrast agents coupled to cell transport vehicles. Mol Img 5, 485–497, 2006.

H.Y. Meltzer et al. Cognition, schizophrenia, and the atypical antipsychotic drugs Proc Natl Aad Sci, 1999.

D. Mendeleyev. The dependence between the properties of the atomic weights of the elements, which described elements according to both weight and valence. Russian Chem Soc, 1869.

S.L. Merbs. J. Nathans. Molecular determinants of human red/green color discrimination. Neuron, 1992.

L. von Mises. Human Action: A Treatise on Economics, 1912.

M. Mishina et al. Function of sigma1 receptors in Parkinson's disease. Acta Neurol Scand, 2005.

C. Moorhead, New York Times, September 16, 1983.

Muna A-Fuzai.K uwaitT imes, April2 007.

W.G. Myers, H.C. Vanderleeden, Radioiodine, I-125. The Ohio State University Health Center, Columbus.

Naigenics and 23andMe. Venture Beat Life Sciences, 2007.

H. Neunhoeffer, P.F. Wiley. The chemistry of heterocyclic compounds, 1978.

A. Newberg. J Nuc Med, 2006.

A. Newberg et al. Regional changes in glucose accumulation during meditation. J Nucl Med, 2006.

R. Niebuhr. Moral Man and Immoral Society, Charles Scribner, 1934.

R. Niebuhr. Theologian of Public Life, 1991.

K. Nilsson. The monoamine oxidase A (MAO-A) gene, family function and maltreatment as predictors of destructive behaviour during male adolescent alcohol consumption. Alcohol Rep, 2007.

K. Nilsson et al. Role of monoamine oxidase A genotype and psychosocial factors in male adolescent criminal activity. Biol Psychiatry, 2000.

Nobel Prize in Medicine was awarded to Daniel Nathans, Werner Arber and Hamilton Smith, 1978.

E.A. Nofzingera et al. Forebrain activation in REM sleep: an FDG PET study. Brain Res, 1997.

M. Odent. How Aphrodite, Buddha and Jesus developed their capacity to love. Midwifery Today, 2001.

J.R. Oppenheimer. Speecht ot he Associationo fL os AlamosS cientists. Los Alamos, NM, 1945

Y. Ouchi et al. Effect of simple motor performance on regional dopamine release in the striatum in parkinson disease patients and healthy subjects: a positron emission tomography study. J Cerebral Blood Flow Metabo, 2002.

F. Ovsiew, Neuropsychiatric approach to the patient. In: B.J. Sadock, V.A. Sadock, editors. Comprehensive Textbooko fP sychiatry. Lippincott Williamsa nd Wilkins,Philadelphia, 2005.

N. Pankratz, T. Foroud, et al. Mutations in LRRK2. Movement Disorders, 2006.

D.C. Park et al. Models of visuospatial and verbal memory across the adult life span. Psychol Aging. 17, 299–320, 2002.

A.M. Paul. Dirac was awarded the Nobel Prize in Physics, 1933.

L. Pauling. The Nature of the Chemical Bond, 1939.

L. Pauling. Sickle Cell Anemia, a Molecular Disease. Science, 1949.

S. Peele. The Meaning of Addiction: Compulsive Experience and Its Interpretation, 1985.

S. Peele, M. Grant. Alcohol and Pleasure: A Health Perspective. Psychology Press, 1999.

M. Phelps. PET: Basic Science and Clinical Practice. Springer, 1975

V. Pierre, R. de Broglie was awarded the Nobel Prize in Physics, 1929.

M. Pines. H. Berger, The Brain Changers: Scientists and the New Mind Control.

S. Pinker. The Moral Instinct. New York Times Magazine, January 13, 2008.

T. Pitman. The Chemistry of Love, 2007.

K.R. Popper. Science as Falsification, 1963.

T. Porkka-Heiskanen et al. Adenosine: a mediator of the sleep-inducing effects of prolonged wakefulness. Science,1 997.

J. Priestley. Experiments and observations on different kinds of air, 1774–1786.

T. De Quincey. Confessions of an English opium eater, 1822.

T. Ramadan. Quoted in NY Times Book Review, 2008

J.L. Rapoport et al. Progressive cortical change during adolescence in childhood-onset schizophrenia. A longitudinal magnetic resonance imaging study. Arch Gen Psychiatry, 1999.

Raymond Bernard Cattell. The Scientific Analysis of Personality, 1977.

J. Reynolds. A Discourse to the Students of the Royal Academy, 1772.

J. Reynolds. National Gallery of Ireland, Dublin, Oliver Goldsmith, 1772.

J.O. Rhine. Brain acetylcholinesterase activity in mild cognitive impairment and early Alzheimer's disease. J Neurol Neurosurg Psychiatry 2003.

L. Richardson. What Terrorists Want: Understanding The Enemy, Containing the Threat, Random House, 2006.

Robertseta l. PharmacolB iochemB ehav 6, 615–620, 1977.

W.C. Röntgen. The Nobel Prize in Physics, 1901.

B.L. Roth, H.Y. Meltzer. The Role of Serotonin in Schizophrenia, Neuropsychopharmacology.

S. Roy. Relationshipb etweenr egionalb loodfl owa ndn euronala ctivity. JP hysiol 11, 85–108, 1890.

E.R utherford.R adioactivity,1 904.

C. Sagan. Broca'sB rain. RandomH ouse, New York, 1979.

G.J.E Schmitt. The striatal dopamine transporter in first-episode, drug-naive schizophrenic patients: evaluation by the new SPECT-ligand[99mTc]TRODAT-1. J Psychopharmacol, 2005.

R. Schoenheimer. The Use of Isotope Tracers to Study Intermediary Metabolism, 1935.

A. Schubiger. Positron emission tomography (PET) to study the role of adenosine A1 receptors in human sleep regulationess. Center of Radiopharmaceutical Science, Paul Scherrer Institute. University of Zurich

W.S chultz.N euroscientist,2 001.

R. Schwarcz, Creese, J. Coyle, S. Snyder. Dopamine receptors localised on cerebral cortical afferents to rat corpus striatum. Nature, 1978.

G. Seaborg was awarded the Nobel Prize in Chemistry, 1961.

D. Selkoe. Alzheimer's disease: arthritis of the brain? – Billions of dollars have been poured into the search for a cure for Alzheimer's disease. New Scientist, 1993.

H. Selye, General adaptation syndrome. Nature, 1936.

M. Shermer. The Mind of the Market, 2008.

C. Sherrington. The Nobel Prize in Physiology or Medicine, 1932.

A.K. Shetty. Neurosci Biobehav Rev, 2007.

C. Shuster. The Infinite Mind: Psychedelics, 2003.

A.B. Singleton. Alpha-synuclein locus triplication causes Parkinson's disease. Science, 2003.

D.H. Small, R. Cappai. Alois Alzheimer and Alzheimer's disease: a centennial perspective. J Neurochem, 2006.

F. Soddy. The Nobel Prize in Chemistry, 1921

L. Sokoloff. FDG in epilepsy. Acta Neurol Scand, 1979.

P.H. Soloff, K.G. Lynch. Characteristics of suicide attempts of patients with major depressive episode and borderlinep ersonalityd isorder. AmJ P sychiatry, 157, 601–608, 2000.

P. Sorokin. Social and Cultural Dynamics, Boston, 1957.

R.W. Sperry. D.H. Hubel, T.N. Wiesel. Nobel Prize in Medicine, 1981

Stockmeir et al. Suicide and Serotonin, 1998.

Suhara et al. Age related changes in human muscarinic acetylcholine receptors examined with positron emission tomography. Neurosci Lett, 1993.

Taminga and colleagues. Arch Psychiatry, 1992

C.A. Tamminga. Limbic system abnormalities identified in schizophrenia using positron emission tomography with fluorodeoxyglucose and neocortical alterations with deficit syndrome. Arch Psychiatry, 1992.

L. Tancredi. Hardwired Behavior, Cambridge University Press, 2005.

R. Tanzi, J. Gusella, in B.A. Yankner, Unravelling a tangled mind. Nature 408, 2000.

M.B. Terry et al. Nature, 1995.

M. Thase. Third International Conference on Bipolar Disorder, 1999.

L. Thomas. The Youngest Science. Viking Press, 1983.

M. Thomas et al. Neural basis of alertness and cognitive performance impairments during sleepiness. I. Effects of 24 h of sleep deprivation on waking human regional brain activity. J Sleep Res, 2000.

J. Thompson. Embryonic Stem Cell Lines Derived from Human Blastocyes Science, 1998.

J. Tiihonen. J Nucl Med, 1997.

J. Tiihonen et al. Eur J Nucl Med 1997.

J. Tiihonen et al. Striatal dopamine transporter density in major depression. J Psychopharmacol, 1999.

J. Tiihonen et al. Antidepressants and the risk of suicide, attempted suicide, and overall mortality in a nationwide cohort. Arch Gen Psychiatry, 2006.

D.E. Tillitt. The Price of Prohibition. Harcourt, Brace and Co., 1932.

R. Torricelli. In our Own Words, Pocket Books, 1999.

D.W. Townsend. A combined PET/CT scanner. J Nucl Med, 1998.

D.W. Townsend. A Combined PET/CT Scanner. J Nucl Med, 1998.

L. Tzu. Tao Te Ching (The Way and Its Power).

I. Urbana. NY Times, 2002.

Vigneaud won the Nobel Prize in

N. Volkow. Curr Opin Neurol, 2007.

N.D. Volkow et al. Therapeutic doses of oral methylphenidate significantly increase extracellular dopamine in the human brain. J Neurosci, 2001.

B. Voytek. Differences in regional brain metabolism associated with marijuana abuse in methamphetamine abusers. Synapse, 2005.

B. Voytek, D.G. Cerebral. Metabolic dysfunction and impaired vigilance. Biol Psychiatry, 2005.

B. Voytek et al. Differences in regional brain metabolism associated with marijuana abuse in methamphetamine abusers. Interscience, 2005.

N. Wade. New York Times, March 20, 2007.

H.N. Wagner Jr. Clinical PET: Its Time Has Come. J Nucl Med, 1991.

H.N. Wagner Jr. A Personal History of Nuclear Medicine. Springer, 2006.

H.N. Wagner Jr. et al. Imaging dopamine receptors in the human brain by positron tomography. Science, 1983.

J.K. Walsh. Chronic Insomnia, Ability to Function and Quality of Life. J o Clin Psychopharmacol, 1991.

C. Walsh, G. Woods, cited by W.J. Cromie. Genes found that regulate brain size: One increases, the other decreases. Harvard Gazette Archives. Yerkes, Hammock, and Young. Nature, 17, 2004

S.S. Wang, Wall Street Journal, July 3, 2007.

T.G. Whittle, R. Wieland. The Story Behind Prozac… the Killer Drug. Church of Scientology.

R.W. Wilkens.P residential Address,C irculation,1 958.

R. Wilkowski. The cognitive basis of trait anger and reactive aggression: An integrative analysis. Soc Psychol Rev,2 008.

G. Wills.S t. Augustine,L ippier/Viking,1 999.

R.R. Wise.C ommentary,2 008.

D.F. Wong, H.N. Wagner Jr, R.F. Dannals, et al. Effects of age on dopamine and serotonin receptors measured by positron emission tomography in the living human brain. Science, 1984.

R.M. Wood et al. Neuropsychopharmacology, 2006.

R.M. Wood. et al. Neuropharmacology, 2006.

D.W. Wooley, E.A. Shaw. Pharmacology and Therapeutics, Elsevier, 2007.

R. Wright.T heM oral Animal,1 994.

J. Wu. Great Thinkers of the Eastern World, Harper Collins, 1995.

T. Wyss-Coray. Blood test a step to predicting Alzheimer's risk?, 2007.

H.H. Xue. Molecular Psychiatry, 2003.

Yerkes,H ammock, Young.N ature,2 004

L.A. Zadeh. Computing with Words in Information/intelligent Systems, 2000.

J. Zeber. Racial Disparities Found in Pinpointing Mental Illness. quoted by Shankar Vedantam in Washington PostJ une2 8,2 005.

Index